Engineering Analytics

Engineering Analysis

Engineering Analytics
Advances in Research and Applications

Edited by
Luis Rabelo, Edgar Gutierrez-Franco, Alfonso Sarmiento,
and Christopher Mejía-Argueta

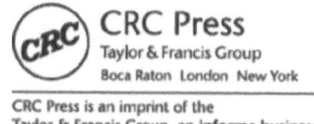

CRC Press
Taylor & Francis Group
Boca Raton London New York

CRC Press is an imprint of the
Taylor & Francis Group, an **informa** business

First edition published 2022
by CRC Press
6000 Broken Sound Parkway NW, Suite 300, Boca Raton, FL 33487-2742

and by CRC Press
2 Park Square, Milton Park, Abingdon, Oxon OX14 4RN

Library of Congress Cataloging-in-Publication Data
Names: Rabelo Mendizabal, Luis C. (Luis Carlos), 1960– editor. |
Gutierrez-Franco, Edgar, editor. | Sarmiento, Alfonso, editor. |
Mejía-Argueta, Christopher, editor.
Title: Engineering analytics : advances in research and applications / edited by Luis Rabelo,
Edgar Gutierrez-Franco, Alfonso Sarmiento, and Christopher Mejía-Argueta.
Description: First edition. | Boca Raton : CRC Press, 2022. |
Includes bibliographical references and index.
Identifiers: LCCN 2021016055 (print) | LCCN 2021016056 (ebook) |
ISBN 9780367685348 (hbk) | ISBN 9780367685379 (pbk) |
ISBN 9781003137993 (ebk)
Subjects: LCSH: Industrial engineering–Data processing. | System analysis–
Data processing. | Big data. | Engineering mathematics.
Classification: LCC T57.5 .E54 2022 (print) |
LCC T57.5 (ebook) | DDC 620.00285–dc23
LC record available at https://lccn.loc.gov/2021016055
LC ebook record available at https://lccn.loc.gov/2021016056

ISBN: 978-0-367-68534-8 (hbk)
ISBN: 978-0-367-68537-9 (pbk)
ISBN: 978-1-003-13799-3 (ebk)

DOI: 10.1201/9781003137993

Typeset in Times New Roman
by Newgen Publishing UK

Contents

Chapter 15 Analysis of Internet of Things Implementations Using
Agent-based Modeling: Two Case Studies...................................255

*Mohammed Basingab, Khalid Nagadi, Atif Shahzad, and
Ghada Elnaggar*

Preface

Engineering Analytics is beginning to consolidate as a set of mandatory tools to learn and obtain vast benefits. Engineering Analytics results from a convergence of advances that have motivated a *tsunami* of applications and successes. Computing has evolved and continues to grow in capacity and speed, and Artificial Intelligence models and Big Data are already maturing, along with advanced statistics, simulation, and operations research.

This book is very innovative and delivers the portfolio of techniques and areas that make up Engineering Analytics, but at the same time, describes a diversified set of sophisticated applications. It provides a perspective on Engineering Analytics and its applications from the descriptive, the visual, the predictive, and the prescriptive. The areas covered in these selected articles include a wide range of applications from the supply chain to the Internet of Things. Techniques include neural networks, Deep Learning, mixed-integer programming, hybrid simulation, and stochastic optimization.

SYNOPSIS OF *ENGINEERING ANALYTICS*

Beginning with a methodological framework, authors Janjevic and Winkenbach in Chapter 1 explore the role of interactive visualization and human-in-the-loop optimization to improve human decision making in the context of supply chain design. The authors elaborate a framework allowing the conceptualization of group decision making in data-driven design solutions that are better capable of responding to many companies' real-life supply chain challenges.

Chapters 2, 3, and 4 cover descriptive and visual analytics. In Chapter 2, authors Clavijo-Buritica, Abushaega, González, Amorim, and Polo present an analysis based on unsupervised learning to find associations and uncovered structures from data of road closures in Colombia. The descriptive study identifies the vulnerable areas and measures the effects of such disruptions on Colombia's performance when the materials' flows are restricted due to landslides or road maintenances. In Chapter 3, authors Prato, Rodriguez, Martínez, Sarmiento, and Talero use Colombia's National Registry of Cargo Dispatches to present a comprehensive study of the need for analytics in order to promote connectivity with transportation networks and foreign trade nodes in Latin America. Analytics will facilitate the development of strategies to improve the logistic capacities for freight transportation.

In the twenty-first century, organizations have increased their expertise in capturing, collecting, and storing data from their customers, operations, employees, and competitors. However, having data is not enough. In Chapter 4, the author Torres deals with this issue, stating that data is essential, and the recent developments in analytics can support the utilization to gain a competitive advantage.

Chapters 5 to 10 concern predictive analytics. Using Blockchain as one example of new technology where there are no clear guidelines to assess its benefits in complex systems such as supply chains, the authors of Chapter 5, Obeidat and Rabelo, introduce a framework to assess the effectiveness of Blockchain using agent-based simulation and Deep Learning. This framework is well explained using a Peer2Peer case study.

Regarding market behavior analysis and product demand prediction, the authors Prada and Ortiz in Chapter 6 discuss using hybrid modeling to estimate the demand of cars, utility vehicles, and pick-up trucks in a megacity (Bogotá, Colombia). The hybrid predictive analytics model allows estimating the demand under different scenarios and responds to a population of consumers with specific socio-demographic characteristics in an established macro-economic context. This is an excellent example of how the power of system dynamics is complemented by agent-based simulation.

The expanding significance of seaports in global trade makes them indispensable for business efficiency and supply chains. In Chapter 7, authors Toukan, Ling Chan, Mejía-Argueta, and Kazemi develop a system dynamics framework that assesses the impact of policy and investment decisions on the container transport chain. The simulation model supports decision makers considering the implications of different transport chain strategies under different scenarios.

Management of data during a pandemic is essential for monitoring and control tasks: the prediction of their behavior becomes fundamental for the planning of actions and management of the crisis. In Chapter 8, authors Manrique Ruiz and Puentes discuss the challenges of data analytics for the novel coronavirus. This chapter is focused on data management and its predictive analysis. Their analysis finds that the selected protocols for data collection, filing, and release lead to constantly growing datasets that require extensive cleansing before the data is used for analysis. The authors use clustering methods to correct the dataset and reduce the health institutions' manual errors. Using the normalized information, a transition matrix between states inside the health system depicts the virus' modeled behavior in the population. Finally, the authors delve into the importance of standardizing and releasing the data correctly.

In Chapter 9, authors Halabi-Echeverry, Obregón-Neira, Niño-Vergara, Aldana-Bernal, and Baron-Perico present in detail the importance of seaport operations and the respective difficulties due to their complexities. Agent-based simulations with the agents' design following standardized scripts (i.e., IEEE-FIPA and BSPL) can support decision-making schemes to solve these problems.

Simulation has traditionally been used in manufacturing and production systems. In Chapter 10, authors Gutierrez, Clavijo-Buritica, and Rabelo propose a framework for integrating dynamic programming, discrete simulation, and reinforcement learning that is described to take advantage of self-adjusting simulation procedures to meet goals for programming purposes in a manufacturing environment. The framework contributes by understanding the decision process for scheduling planning and integrating analytical techniques to learn effective scheduling policies.

Chapters 11 to 14 cover prescriptive analytics. In Chapter 11, the author Padilla presents an integrated support proposal for sales and operations planning. On the demand side, it includes using a Random Forest Regressor to forecast sales in terms of average prices. The supply side uses a mixed-binary programming model that maximizes the present value of the monthly flow of marginal contributions, including supply and demand decisions.

In Chapter 12, authors Garcia-Bedoya, Ferreira, Clavijo-Buritica, Gutierrez-Franco, and Lowe present a software architecture for autonomous vehicles which requires infrastructure in the cloud and the use of convolutional network algorithms with different layers. The main findings are that Deep Learning can create advanced models that can generalize with relatively small datasets. These tasks are distributed in different hardware to integrate the reliability of automotive embedded systems with Linux architecture's flexibility.

The coffee industry has grown to be one of the world's biggest industries, and in Chapter 13 the authors Botero, Salman, and Tayaksi present a mixed-integer linear programming network model that determines the most cost-efficient coffee supply chain network configuration. The modeling considers farmers from Caldas, Colombia, to the U.S. Northeastern region. This study presents cost reduction opportunities for coffee farmers and opens a door for improving their livelihoods. Chapter 14 also discusses food and retail operations, asserting that accessibility plays a significant role in food security in Colombia. The authors da Silva, Granados-Rivera, Mejía, Mejía-Argueta, and Jarrín use geographical information systems to develop geographical clusters. These clusters identify how the physical distribution in different distribution channels or retailers can influence the proximity of fresh food at retailers in the different areas of the Sabana Centro region of Colombia. They conclude that different channels ensure food accessibility to diverse groups. They show that small, family-owned retailers or *nanostores* currently play a significant role in making items available for vulnerable population segments, mainly in areas farther away from downtown areas.

Applications in different areas of the industry of intelligent systems and enabling technologies, such as the Internet of Things (IoT), are growing enormously in all societies. In Chapter 15, authors Basingab, Nagadi, Shahzad, and Elnaggar present two case studies to examine these trends. The first study, concerning maintenance activities, models a failure rate control parameter in different scenarios, while orders to pharmacies and the warehouse act as the main active agents. In the second case study, the maintenance activity of a medical facility's IT rooms is modeled using IoT to reduce the overall cost and to analyze the effects of employing IoT.

ACKNOWLEDGMENTS

We would like to acknowledge all the people and organizations that contributed in one way or another to the achievement of this compendium of research and applied cases in Engineering Analytics. First, we would like to express our sincere thanks to the authors of the chapters for sharing their research with the academic community and practitioners in Engineering Analytics and their time, commitment, and speed in 2020. Our thanks to Taylor & Francis for publishing this book, along with their guidance and patience. We are confident that this book contributes to the Analytical Engineering community, especially those in Decision Science, Operations Research, Simulation, and Machine Learning. We hope this book will inspire continued research and development in Engineering Analytics and serve as a foundation for present and future generations of business and data analytics engineers.

Editor Biographies

Luis Rabelo, Ph.D., was the NASA EPSCoR Agency Project Manager and is currently a Professor in the Department of Industrial Engineering and Management Systems at the University of Central Florida. He received dual degrees in Electrical and Mechanical Engineering from the Technological University of Panama and Master's degrees from the Florida Institute of Technology in Electrical Engineering (1987) and the University of Missouri-Rolla in Engineering Management (1988). He earned a Ph.D. in Engineering Management from the University of Missouri-Rolla in 1990, where he also undertook Post-Doctoral work in Nuclear Engineering in 1990–1991. In addition, he holds a dual MS degree in Systems Engineering and Management from the Massachusetts Institute of Technology (MIT). He has completed over 280 publications, has three international patents being utilized in the aerospace industry, and has graduated 40 Master and 34 Doctoral students as advisor/co-advisor.

Edgar Gutierrez-Franco is a Postdoctoral Associate at the Massachusetts Institute of Technology, at the Center for Transportation and Logistics. He also serves, since 2009, as a Research Affiliate at the Center for Latin America Logistics Innovation (CLI) as part of the MIT Global SCALE network and is Fulbright Scholar since 2014. He has experience in consultancy, retail, and beverage industry (LOGYCA, Falabella, and SABMiller in Latin America). Also, during his stay at the Center for Transportation and Logistics, Massachusetts Institute of Technology (2009–2010), he participated in projects concerning Supply Chain Innovation in Emerging Markets (city logistics) and Carbon-Efficient Supply Chains/Sustainability. His educational background includes a B.S. in Industrial Engineering from the University of La Sabana (2004, Colombia), an M.Sc. in Industrial Engineering from the University of Los Andes (2008, Colombia), and a Ph.D. in Industrial Engineering and Management Systems from the University of Central Florida. His expertise includes the use of Machine Learning, operation research, and simulation techniques for supply chain management, systems modeling, and optimization.

Alfonso Sarmiento is an Associate Professor at the Program of Industrial Engineering, University of La Sabana, Colombia. He completed his Bachelor's degree in Industrial Engineering from the University of Lima, Perú. He earned an M.S. from the Department of Industrial and Systems Engineering at the University of Florida, and a Ph.D. in Industrial Engineering with emphasis in Simulation Modeling from the University of Central Florida. Prior to working in academia, Dr. Sarmiento had more than 10 years' experience as a consultant in operations process improvement. His current research focuses on supply chain stabilization methods, hybrid simulation, and reinforcement learning.

Christopher Mejía-Argueta is a Research Scientist at the MIT Center for Transportation and Logistics. He is Director and founder of the MIT Food and Retail Operations Lab (FaROL) where he develops applied research to improve the performance of supply chains. He is also director of the MIT SCALE network for Latin America and the Caribbean, as well as the MIT GCLOG program. He has more than 13 years of experience in analyzing supply chains, and has solved logistical problems for more than 15 countries on four different continents. He is the editor and author of four books on supply chain management for emerging markets and has published dozens of industrial projects. He holds a Ph.D. and a M.Sc. from Monterrey Tech, Mexico.

Introduction

Currently, data analytics and mathematical models are used to facilitate decision making for industrial and service operations worldwide. The term *analytics* includes statistical techniques, mathematical programming, simulation, operations research, Machine Learning, and Artificial Intelligence (with the ever-growing computational power) to support managerial decisions. Therefore, analytics has become a fundamental tool for decision making in different organizational dimensions at multiple complexity levels. Data and Engineering Analytics has become a first step to create knowledge and make better decisions through various problems of organizations and supply chains. Its use has demonstrated the construction of competitive advantage and Corporate Social Responsibility. Its practice creates sustainable organizations, benefits society, and is valuable to businesses (Porter and Kramer 2007).

The impact of analytics has boosted other management initiatives like lean six-sigma, customer segmentation for resource optimization, pattern identification, classification strategies, and forecasting. For instance, optimal inventory management, logistic control towers, analysis and forecasting of sales and operations for higher revenues and customer satisfaction, and consumer behavior modeling are areas being impacted by analytics (Davenport 2006). These areas are vital in creating value for business ecosystems, end consumers and beneficiaries, non-government organizations (NGOs), and other relief operations. In general, analytics is divided into four main areas:

1. **Descriptive Analytics:** Here, descriptive statistics and data mining are commonly used to achieve segmentation, dimensionality reduction via spectral techniques, and classification. Generally, large amounts of data are analyzed to discover patterns and trends, understand potential causes and effects, and identify bottlenecks and opportunities.
2. **Visual Analytics:** This implies migrating from static, reactive data analytics to a more dynamic, proactive visualization enabled by dashboards to analyze and derive important conclusions from data. The visual analysis combines descriptive and inferential statistics techniques with specific knowledge in engineering and systems management to show accurate, concise information to capture a system's behavior and understand trends and cycles. The difference concerning

DOI: 10.1201/9781003137993-1

descriptive analytics is based on how information is shown to focus decision makers and capture the system's right picture. The design of dashboards with business intelligence software to show the key performance indicators (KPIs) is the trend across organizations; however, new trends show the need to include testimonials and qualitative information to build a holistic perspective.

3. **Prescriptive Analytics:** This type of analytics oversees the best course of action for a given situation. It implies that decision- and policymakers aim to use their limited assets optimally (or as close as possible to optimality) and create an action plan in complex circumstances. Techniques such as operations research formulations (e.g., mixed-integer programming), dynamic programming, stochastic modeling, and simulations (e.g., discrete-event, system dynamics, agent-based) are widely used. Results from this type of analytics allow finding implications of various policies, operations, and actions in practice without investing in real pilots and making the best use of available resources (e.g., technology, human talent, fleet, equipment, infrastructure).

4. **Predictive Analytics:** Here, classical linear and non-linear regressions, regression trees, Random Forests, and neural networks are the preferred methodologies. This type of analytics implies analyzing historical data to make estimations about future or unknown events. It is used to make several decisions, including but not limited to forecasting, inventory management, customer and traffic behavior, and scenario planning, among others. It is instrumental because decision- and policymakers can build a roadmap of actions among diverse potential scenarios.

Another aspect of this era of growing available information is how organizations handle Big Data issues, particularly for institutions where the information increases each second due to internal and external transactions and human interactions. *Big Data* applies to large, complex datasets that cannot be handled using traditional data management techniques or simple software. A group of technologies deals with Big Data analytics, which refers to the methods and techniques to extract patterns and new information from structured, semi-structured, and/or unstructured data. This type of analytics is usually contained in Descriptive Analytics. Still, given the rising variety and quantity of data, these authors decided to emphasize that traditional ways to analyze them are becoming quickly outdated. We live in a hyperconnected society due to technology accessibility, dynamic consumption behavior, and globalization— Figure I.1 displays how the data flow across different stages throughout different analytics paradigms (Gutierrez Franco 2019).

Once the data are obtained, a process of cleaning, organizing, and storing starts. This process is followed by data analytics and implementation. Despite cleaning, organization and storage are usually not profoundly described in methodologies. These tools help handle data volume and diversity, remove/treat imprecisions, and provide robust solutions. In that sense, data mining supports a preliminary analytics step that allows decision makers and modelers to gain more knowledge about the organization, system, or supply chain and circumstances to activate more in-depth analytics in subsequent steps.

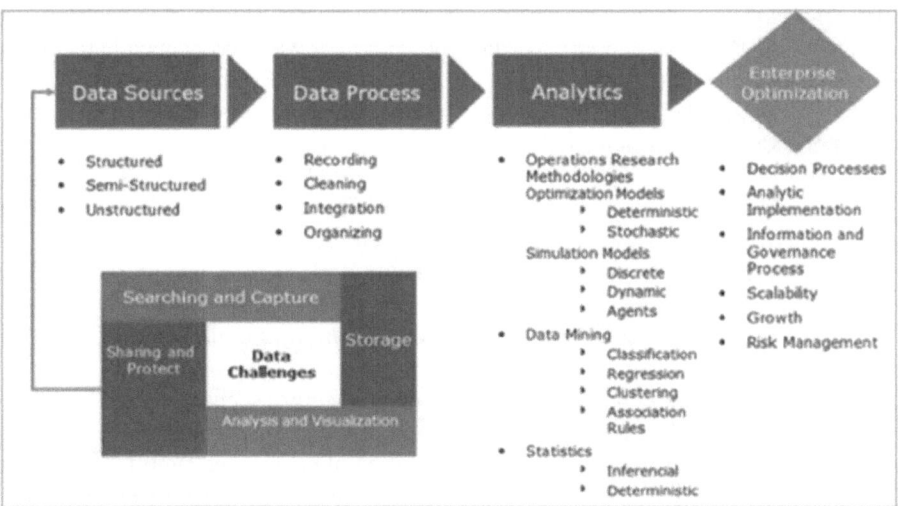

FIGURE I.1   Data-driven enterprise optimization

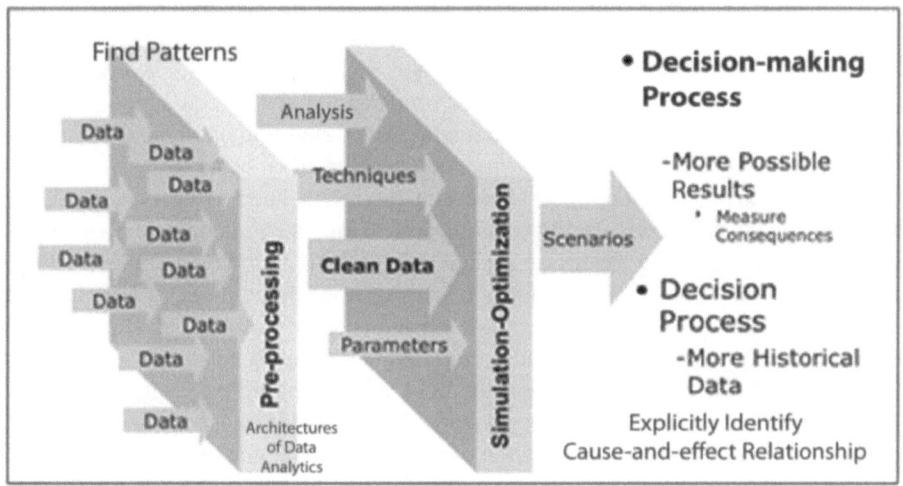

FIGURE I.2   Example of data analytics stages

Data mining and cleansing techniques, together with data analytics, allow the extraction of valuable insights from the data to support the decision-making process. Figure I.2 depicts how the data go through different stages like pre-processing with some statistical and data-mining techniques to prepare for a simulation-optimization. After the data are processed, and the analysis is done, a decision-making process is supported. A series of scenarios to take the decision are listed at the end of the analysis. Furthermore, with this analysis, the data scientists can also discover causalities.

Data analytics is used to find patterns, understand what happened in the past, and have a solid base to apply predictive analytics and infer what can happen in the future. These steps are the basis of any high-performance decision support system.

Considering data analytics, value creation is demonstrated in several cases (Brynjolfsson, Hitt, and Kim 2011):

1. An organization may obtain valuable information regarding its performance in a detailed and convenient manner (e.g., monitoring and visualization).
2. An organization may combine data from multiple sources and utilize resources more efficiently as products and services are targeted to meet specific needs through customizable actions (e.g., optimization, simulation modes, hybrid modeling).
3. An organization may derive a set of potential future scenarios from building a competitive advantage, envisioning new products or markets; as well as find ways to find patterns to address the evolution of the market, suppliers, and other stakeholders (e.g., predictive techniques, forecasting, econometric models).
4. Human force is replaced or supported by algorithms, therefore complementing and augmenting human capabilities (e.g., Artificial Intelligence).

The construction of dynamic business metrics can be achieved throughout the business' detailed analysis (Kaplan 2009). The current techniques, methodologies, and architectures affect how organizations and their supply chain stakeholders can use available data to impact business performance. Analytics can support the construction of these indicators in computational times, enabling companies to improve productivity, visibility, and a long-lasting competitive advantage.

Based on the facts mentioned above and the rising need to analyze significant amounts of data efficiently, engineers now have to meet professional challenges with various tools. This book discusses various engineering analytics methods and their application with case studies in different industries and countries. It is built based on experience from contributors; authors in industry and academia. The goal is to expose the required techniques and methodologies to apply analytics and business intelligence. Readers will learn from the book content based on visual, descriptive, prescriptive, and predictive analytics. Practical examples and case studies support the chapters to demonstrate the application of analytics intelligence.

The content of the book provides introduction to the following fundamental concepts as follows:

- Visual data analytics, Geographical Information Systems, statistics, and probability (Chapters 1, 2, 3, 4, 14)
- Machine Learning (Chapters 2, 10, 12)
- Simulation and optimization models (Chapters 6, 8, 9, 10, 11, 13, 15)
- Strategic issues associated with business data analytics (Chapters 1, 5, 6, 7, 9, 11, 13, 14, 15)
- Hybrid modeling techniques (Chapters 6, 10, 15)
- Trends (Chapter 15)

The information is designed to provide a big picture of analytics using real cases from industry (i.e., automobile, manufacturing, agribusiness, industry sectors) to illustrate the use of multiple techniques. The chapters have different perspectives and provide analytics solution approaches over different case studies from the U.S., Portugal, Colombia, Germany, Jordan, Australia, Peru, and Saudi Arabia. Although this book is not a comprehensive collection of data analytics, it builds upon the need to better link theory in data analytics with practice.

REFERENCES

Brynjolfsson, Erik, Lorin Hitt, and Heekyung Kim. 2011. "Strength in numbers: How does data-driven decisionmaking affect firm performance?" *SSRN Electronic Journal,* 1 (April).

Davenport, Thomas H. 2006. "Competing on analytics." *Harvard Business Review*, January 1, 2006.

Gutierrez Franco, Edgar. 2019. "A Methodology for Data-Driven Decision-Making in Last-Mile Delivery Operations." University of Central Florida. https://stars.library.ucf.edu/etd/6497.

Kaplan, Robert S. 2009. "Conceptual foundations of the balanced scorecard." In *Handbooks of Management Accounting Research*, edited by Christopher S. Chapman, Anthony G. Hopwood, and Michael D. Shields. New York: Elsevier, chapter 3: 1253–1269.

Porter, Michael and Mark Kramer. 2007. "Strategy and society: The link between competitive advantage and corporate social responsibility." *Strategic Direction,* 23 (5).

1 Interactive Visualization to Support Data and Analytics-driven Supply Chain Design Decisions

Milena Janjevic and Matthias Winkenbach

1.1 INTRODUCTION

Supply Chain Design problems typically involve defining the configuration of logistics facilities (i.e., their location, size, service zone) in a manner that maximizes some measure of performance. The Supply Chain Design field is rooted in conventions developed in the 1990s where the focus was on minimizing the total cost of operations, including facility cost, warehousing cost, and transportation cost, among other things. The underlying models used were discrete facility location models solved through various optimization methods (Melo, Nickel, and Saldanha-Da-Gama 2009). The fidelity of these models was constrained by limited data availability and limited computational power.

Today, however, the operating environment of most companies has changed dramatically. Over the last few decades, due to rapid globalization, the proliferation of digital technology and electronic commerce, the growing importance of brands and direct-to-consumer business, and a number of drastic geo-political changes, the conditions under which global supply chains have to efficiently and reliably match supply and demand have become increasingly volatile, uncertain, complex, and ambiguous.

Contemporary Supply Chain Design approaches need to appropriately address these changes in the competitive operating environment of companies. Specifically, they need to make effective use of the abundance of data that is available to companies today, the tremendous advances in predictive and prescriptive analytics, as well as the exponential growth in computational capabilities that recent decades have brought about. Moreover, as service speed, reliability, and flexibility have become critical determinants of competitive advantage, contemporary Supply Chain Design needs to incorporate the non-trivial trade-offs between revenue and cost.

In this chapter, we explore the roles of interactive visualization and human-in-the-loop optimization as a way to improve decision making around Supply Chain Design. These new approaches combine the power of analytical models and the implicit knowledge of expert human decision makers, to improve decision-making transparency and elaborate design solutions that are more capable of responding to the real-life challenges of companies.

DOI: 10.1201/9781003137993-2

1.2   DECISION MAKING IN SUPPLY CHAIN DESIGN

The design of supply chain networks typically entails the selection of a specific network configuration among multiple alternatives based on one or multiple objectives and can therefore be seen as a typical *decision-making problem*. In the following section, we discuss the main characteristics of this problem and the approaches to decision making in this field.

1.2.1   CHARACTERISTICS OF THE SUPPLY CHAIN DESIGN DECISION-MAKING PROBLEM

Supply Chain Design is a decision-making problem with some notable characteristics that render it particularly challenging. First, Supply Chain Design is a *strategic decision-making problem*. It involves strategic decisions on the number, location, capacity, and functions of the production and distribution facilities (Govindan, Fattahi, and Keyvanshokooh 2017). These typically involve substantial investments in infrastructure and are made on a long-term planning horizon. These strategic decisions are interrelated with medium-term tactical decisions (e.g., transportation modes, outsourcing) and short-term operational decisions (e.g., flow of goods in the network). As noted by Klibi, Martel, and Guitouni (2016), Supply Chain Design is a hierarchical decision-making problem, because to make the strategic decisions, decisions at lower levels must be anticipated.

Second, Supply Chain Design is a *complex decision-making problem*. Here, we employ this term as it is defined in the field of complexity theory (see, e.g., Choi, Dooley, and Rungtusanatham 2001). As noted by Pathak et al. (2007), supply networks typically involve complex interconnections between multiple suppliers, manufacturers, assemblers, distributors, and retailers. In addition, planners are faced with the inability to fully anticipate operational and tactical decisions (Klibi, Martel, and Guitouni 2016). The inability to completely describe the system in terms of its individual constituents and to entirely predict system behavior based on the individual parts results in its complexity (Cilliers and Spurrett 1999; Choi, Dooley, and Rungtusanatham 2001).

Third, Supply Chain Design is a *group decision-making problem*. Given the strategic role of Supply Chain Design for corporate strategy, choices in this area are relevant to multiple objectives and stakeholders. For example, as demonstrated by Lim and Winkenbach (2019), the configuration of last-mile supply chain networks is directly linked to service levels proposed by the company and has a major impact on the width of its product assortments. These decisions cannot therefore be taken in isolation by the supply chain department but require input and cooperation with other corporate functions, such as sales, marketing, and finance.

1.2.2   DECISION MAKING IN THE CONTEXT OF SUPPLY CHAIN DESIGN

Decision making has been explored from a number of perspectives. While there is no definite theoretical model on different steps and processes involved in decision making, we can differentiate between two main categories of decision making: *analytical decision making* and *intuitive decision making* (Cohn et al. 2013). Analytical decision making is often associated with "rational" models of decision making and is characterized by systematic and deliberate information analysis in order to reach a decision (Cohn et al. 2013; Dane and Pratt 2007). Conversely, intuitive decision making is characterized by rapid, non-conscious, and holistic associations, with decisions resulting from the interactions between prior knowledge and experience and the new incoming information (Dane and Pratt 2007; Cohn et al. 2013; Evans 2008).

The use of optimization models to support decision making in Supply Chain Design supports the analytical approach. The increase in the computational power and advances in optimization methods over time have significantly increased the ability of these models to accurately represent real-life business context (e.g., integrating uncertainty through stochastic optimization). However, no matter how elaborate, models are inherently limited and unable to comprehensively represent all aspects of a given problem (e.g., accounting for traffic congestion when estimating a delivery lead time). While increasing model fidelity improves their accuracy, it also entails higher complexity and can result in misunderstanding and lack of trust in what is perceived to be a "black-box" process (Meignan et al. 2015).

Intuitive decision making often translates to choices based on previous experience and implicit knowledge. Research on intuition effectiveness finds that this approach can be superior to analytical approaches, especially in complex decision situations where the amount of information is too huge to be processed deliberately Julmi (2019). However, it can also lead to a number of well-documented cognitive biases, such as the anchoring effect, the shared information bias, or the selection of the "satisficing" rather than the optimal solution (see Carter, Kaufmann, and Michel (2007) for a comprehensive review of judgement and decision-making biases in supply chain management). Literature shows that optimal strategic decision making may require both rationality and intuition (Calabretta, Gemser, and Wijnberg 2017). In line with this finding, Supply Chain Design can benefit from both approaches.

In addition, the multi-stakeholder nature of Supply Chain Design decision making requires approaches that are applicable in group settings where different team members, with potentially different knowledge and objectives, must collaborate to reach a decision. Here, the rational model of decision making generally prescribes decision-support tools integrating the objectives of different stakeholders, and where optimal trade-offs are found through techniques such as multi-objective optimization models or multi-criteria multi-actor analysis. However, in complex decision making characterizing with incomplete information and ambiguous objectives, limitations of pure modeling approaches have been acknowledged (see, e.g., Lebeau et al. 2018; Le Pira et al. 2017). Increasingly, the decision making and operational research literature suggest approaches promoting shared situational awareness and joint knowledge production with the aim of reaching consensus (see, e.g., Janjevic, Knoppen, and Winkenbach 2019; Hegger et al. 2012).

1.2.3 NEW PERSPECTIVES ON DECISION MAKING IN SUPPLY CHAIN DESIGN

We identify two potential avenues for increasing the effectiveness of decision making in Supply Chain Design problems. The first avenue relates to methods that aim to combine analytical and intuitive approaches to decision making. Here, we focus on the approaches: *human-in-the-loop optimization* methods and *visual analytic*. The second avenue relates to the facilitation of *group decision-making processes*. These various approaches are now detailed.

Augmenting model-based decision making through human-in-the-loop optimization. Supply Chain Design is a semi-structured problem: it entails structured components (e.g., ability to model the network and employ data to quantify its performance) and unstructured components (e.g., ambiguous objectives or incomplete knowledge). Solving semi-structured problems involves a combination of both standard optimized solution procedures and human intuition or judgments (Niu, Lu, and Zhang 2009).

In the field of Supply Chain Design, human-in-the-loop optimization aims to leverage the combined strengths of machines and humans. A number of techniques can be employed to this end, such as interactive (multi-objective) re-optimization, interactive evolutionary algorithms, and human-guided search (see Meignan et al. (2015) for a comprehensive review). Different techniques allow for refinement of model specifications in an iterative fashion and/or fine-tuning of solutions which in turn enables users to gain better intuition about model results.

The interaction between the human and the optimization algorithm is typically established through a visual interface. However, to the best of our knowledge, literature still lacks a discussion on how to integrate human-in-the-loop optimization and visual analytic to enhance decision making. Liu et al. (2017) explore relationships between interactive optimization and visualization and remark that most extant literature in the field is almost completely silent about visualization and interaction techniques and the user experience. Extant human-in-the-loop optimization research is mainly focused on the development of efficient algorithms and is almost completely silent on the relationships between their use and behavioral processes involved in decision making.

Decision support through information visualization. Information visualizations can amplify human cognition by transferring strenuous cognitive operations with abstract data into visual reasoning processes with external graphic representations (Windhager, Schreder, and Mayr 2019). Visual analytics combines automatic analysis with interaction and visualization techniques to gain knowledge from data (Keim et al. 2010). Several contributions (see, e.g., Kohlhammer 2005; Kohlhammer, May, and Hoffmann 2009) address information visualization from a decision-making perspective. Here, the objective is not to replace human decision making but to enhance decision-making abilities for the user by providing context-sensitive visualization (Kohlhammer, May, and Hoffmann 2009). Visualization becomes the medium of a semi-automated analytical process, where humans and machines cooperate using their respective distinct capabilities for the most effective results (Kohlhammer, May, and Hoffmann 2009).

Visual methods are used to collaborate in the construction of shared knowledge in many fields (Kirschner, Buckingham-Shum, and Carr 2012). Information visualization is found to support group decision-making processes by supporting social interaction and sense making, discussion, and consensus building (Heer and Agrawala 2008). It can be particularly relevant when addressing complex group decision-making problems. For example, geo-visualization tools are increasingly used to support spatial planning processes characterized by complex geographic space, multiple actors with different roles, and tacit criteria and knowledge (Andrienko et al. 2007; Lami and Franco 2016).

Group decision making in technology-supported decision making. The necessity of creating shared situational awareness and joint knowledge production in group decision-making settings is increasingly acknowledged. In model-driven group decision making, the field of behavioral operational research emphasizes the role of operational research models in conceptualizing a shared understanding of the problem across team members (Franco 2013). Similarly, multi-attribute decision-making literature (see, e.g., Dong et al. 2016; Janjevic, Knoppen, and Winkenbach 2019) uses multi-attribute decision-making models to create this shared understanding. In both approaches, decision making is decomposed in a number of phases. Multiple iterations or decision-making cycles can be performed in order to reach consensus. These approaches are typically enabled through external moderation, which is in line with the idea of the neutral facilitator present in the consensus-building literature (see, e.g., Innes and Booher 1999). Our approach focuses on employing technology to enable effective collaboration between human decision makers and between human decision makers and machines. To investigate the impact of technology on group decision making, we draw from findings of macro-cognitive research.

Macro-cognitive theory is a multi-disciplinary area of research focused on understanding how groups, teams, and other collective entities learn, develop meaningful knowledge, and apply it to resolve significant and challenging problems (Kozlowski and Chao 2012). It focuses on decision-making settings characterized by ill-defined goals, and high levels of uncertainty, complexity, and time pressure (Cacciabue and Hollnagel 1995; Klein, Klein, and Klein 2000; Klein et al. 2003). These features are commonly found in organizational strategic decision making around Supply Chain Design.

Foundational to macro-cognition theory is the notion of extended cognition, which argues that the brain is inextricably coupled to one's external environment and often relies on this coupling for many complex tasks (Fiore and Wiltshire 2016). The role of technology has been discussed from the earliest conceptualizations of the macro-cognitive theory (see, e.g., Klein et al. 2003; Klein, Klein, and Klein 2000). In particular, external representations (a concept close to that of "boundary objects" employed in the area of Computer Supported Cooperative Work) are particularly useful for settings where teams are supported by technology and where members deal with a tremendous variety of data and information (Fiore and Wiltshire 2016).

It is generally recognized that technology can support team cognition in two ways. The first type of support is *offloading*, where cognitive work is performed externally in order to free up cognitive resources (Rosen 2010; Fiore and Wiltshire 2016). Cognitive resources are freed up through memory aid or the offloading of the computation to the machine. The second type of support is *scaffolding*, where cognitive processing is not replaced, but its effectiveness is increased (Rosen 2010; Fiore and Wiltshire 2016).

1.2.4 SYNTHESIS

Both human-in-the-loop optimization and visual analytics aim to leverage the complementary strengths of humans and machines to solve difficult real-world problems.

When applied in a group decision-making context, these lead to an interplay between various levels (i.e., individual and team) and different types of cognitive systems (i.e., biological and digital). While decision-making literature and behavioral operations research offer interesting insights on processes occurring in model-supported group decision making, they generally prescribe external human facilitation. Concepts from macro-cognitive theory allow to address settings where such facilitation is performed through technological artifacts.

1.3   INTERACTIVE VISUAL ANALYTICS IN SUPPLY CHAIN DESIGN

In the following section, we present a framework that allows to conceptualize decision making facilitated by such a system. We then present the main precepts of design of such a system.

Based on the findings from the literature, we can differentiate the decision-making process occurring according to three main dimensions, as illustrated in Figure 1.1. The *first dimension* is relevant to the decision-making phase, i.e., problem analysis, model specification, solution generation, and solution evaluation. The *second dimension* is relevant to the type of interactions occurring in the team decision-making process. These can be human-to-human interactions (e.g., discussing opportunities and various Supply Chain Design scenarios) or human-to-machine interactions (e.g., changing model parameters, running the optimization procedures). The *third dimension* is relevant to the role of technology as identified in the macro-cognitive research. In our case, offloading allows to free cognitive resources by enabling storage of a large repository of data and the exploitation of the computational power of machines to run large-scale optimization models. In contrast, scaffolding does not aim to replace cognitive processes in decision making, but to augment it through use of technology. Here, we draw from technology-supported learning research to differentiate between two types of scaffolding: *conceptual scaffolding* and *implicit scaffolding*. Conceptual scaffolding is enabled through visual analytic. It is implemented through various visualizations that seek to guide information interpretation, such as diagrams, matrices, and graphs (Park 2017; Suthers and Hundhausen 2001; Alexander, Bresciani, and Eppler 2015). Implicit scaffolding employs interactive users interfaces and feedback in order to build intuition (Podolefsky, Moore, and Perkins 2013). In our case, it is enabled through human-in-the-loop optimization.

In addition, we can add a temporal dimension to the decision-making process. In line with literature, rather than employing a linear approach, teams are performing multiple decision-making loops where each loop is composed of several decision-making phases and materialized through various technology-supported team interactions (see Figure 1.1). Different teams may employ different decision-making strategies. Some teams might spend more time processing information within each decision-making loop and make greater use of visual analytic to explore data and evaluate results. Others might perform faster, more numerous decision-making loops and make greater use of human-in-loop optimization to explore the problem space and adjust their solutions. Figure 1.2 represents the resulting system design. Users interact with two main components: (1) an interactive geo-spatial interface which

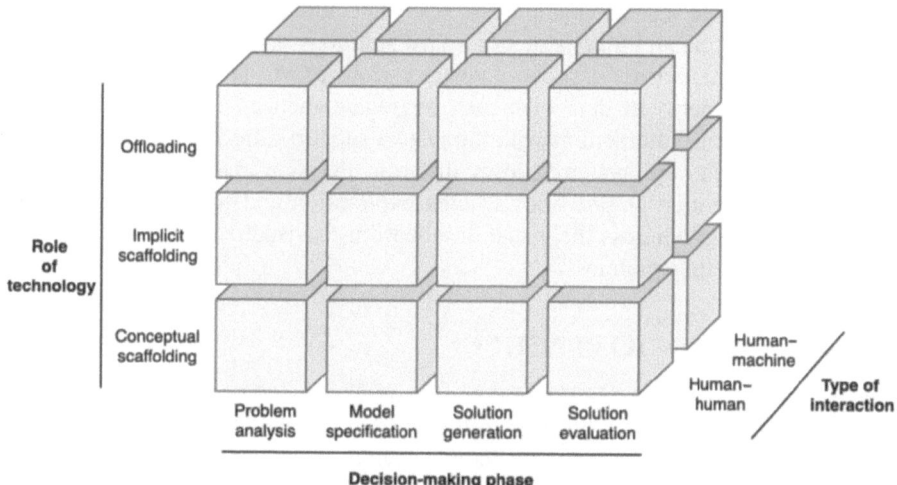

FIGURE 1.1 Decision making supported by interactive visualization: *(left)* dimensions of the process, and *(right)* temporal aspects

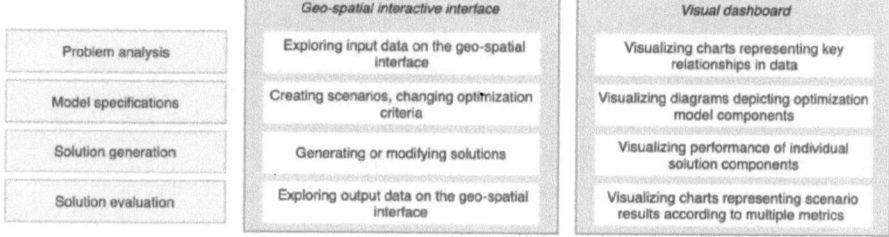

FIGURE 1.2 System design overview

allows teams to input scenario parameters, select and adjust optimization criteria and their weights, and manually construct or adjust solutions proposed by the optimization engine; and (2) a visual dashboard where various charts and diagrams represent key relationships between data elements, aggregate metrics, optimization model structure, etc. Various interactions with the system support different decision-making phases. In addition, the system allows different decision-making strategies: the interactive optimization engine allows quick exploration and feedback, supporting implicit scaffolding; whereas the visual dashboard uses visualizations to support the conceptual scaffolding strategy.

1.4 APPLICATION TO PRACTICE

In the following section, we present two real-world case studies in which the above-described conceptual framework was followed to create interactive visual analytics systems supporting complex Supply Chain Design decisions of major multi-national companies. Specifically, these case studies are based on research collaborations of the MIT CAVE Lab (http://cave.mit.edu) with these companies. The MIT CAVE Lab is a research initiative of the MIT Center for Transportation & Logistics working on a comprehensive software development framework for interactive visual applications that combine the power of analytical models with the tacit domain knowledge of decision makers to allow for efficient, data- and analytics-driven Supply Chain Design decision making.

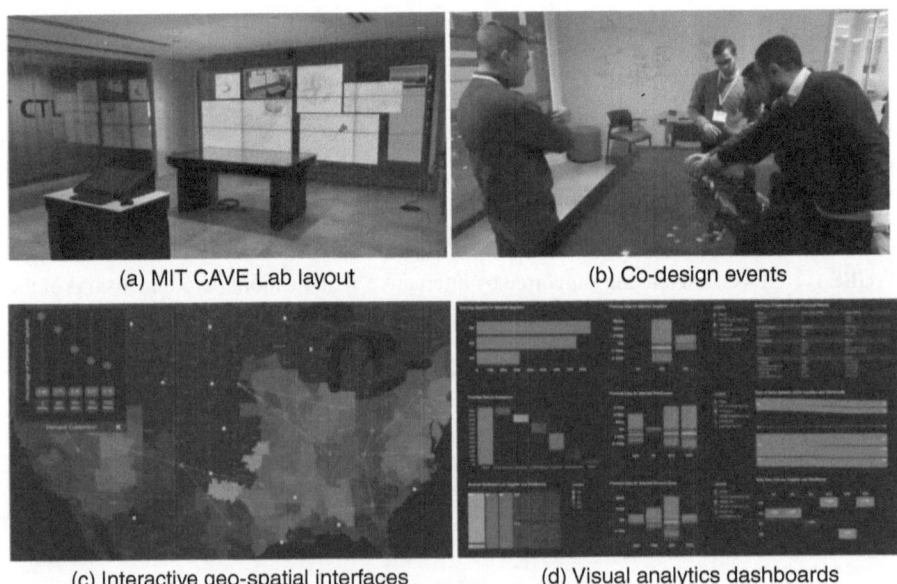

(a) MIT CAVE Lab layout   (b) Co-design events

(c) Interactive geo-spatial interfaces   (d) Visual analytics dashboards

FIGURE 1.3   MIT CAVE Lab: *(top left)* lab layout, *(top right)* co-design event, *(bottom left)* interactive geo-spatial interfaces, and *(bottom right)* visual analytics dashboard

1.4.1 DISTRIBUTION NETWORK DESIGN FOR A MULTI-NATIONAL CHEMICAL COMPANY

Problem setting. A leading chemical company wishes to identify regions in North America in which it can profitably expand its market share by re-designing its distribution network. This network re-design decision is *strategic*, as it involves significant investments into additional network facilities and a fundamental change in the company's transportation flows in this important market. The decision-making problem is also *complex*, as the company needs to consider non-trivial trade-offs between the cost effects and the revenue effects of any potential change in network structure. A densification of the network to increase overall proximity to demand leads to improved service levels, which in turn drives market share and thus revenue. However, it also leads to increased operational cost and the exact customer response to service-level improvements is unknown. Lastly, the problem is also a *group decision-making* problem. Specifically, the cost effects of any network re-design affect the budget and performance metrics of the supply chain division, while the revenue effects predominantly affect business metrics that are relevant to the sales department. Thus, consensus needs to be reached across multiple stakeholders with potentially diverging incentives. Moreover, representatives from different departments and local branches of the company have tacit knowledge about local market conditions that may help in predicting customer response.

Application of visual analytics to support decision making. Following the conceptual framework outline in Section 1.3, we developed a visual analytics system that effectively supports the company in this complex decision-making problem. Based on historical market data, we conducted an *exploratory analysis* to establish a model of market share elasticity in response to changes in service level. We then integrated a non-linear generalized demand model into a mixed-integer linear programming model to optimize the network structure with the objective of optimizing for profit, rather than cost or revenue. We then conceptualized and implemented an *interactive visual interface* that allows users to intuitively explore the sensitivity of the model results to changes in some of the key determining parameters of the model. Specifically, users can explore the sensitivity to changes in the uncertain customer response to increased geographic proximity of supply and demand by making arbitrary changes to the characteristic shape of the non-linear demand function. Users can structure their decision-making process by creating and exploring multiple scenarios and re-optimizing them on demand. The resulting cost, service level, and market share figures obtained from optimizing a scenario are fed back to the user at various levels of visual aggregation, from a geographical mapping on the ZIP code level to facility- and network-level aggregations represented by charts on a dashboard. Lastly, we convened members of the company's logistics, finance, sales, and marketing departments for a day-long *interactive co-design workshop* using the optimization model and visual interface in the MIT CAVE Lab.

Benefits of the approach. The chosen approach yields a number of advantages to the company's Supply Chain Design decision-making problem. On the process side, the combination of sophisticated analytics with an intuitively accessible visual interface enabled faster, more data- and analytics-driven decisions. Besides quantitative data, the decision-making process was able to systematically incorporate the tacit knowledge of domain experts from various departments and business units of the company. In terms of decision effectiveness and barriers to implementation, the chosen decision-making approach led to stronger organizational buy-in across stakeholders. The interactive exploration and discussion of the problem allowed the various stakeholders to reach a common problem understanding and to better align their interests and incentives. In terms of business success, the group decision-making process facilitated by the visual analytics systems we developed led to a bolder re-design of the company's network and enabled a stronger growth in overall market share and profitability than a traditional design approach would have been able to. More specifically, it enabled a lasting shift in the mindset of the relevant decision makers, away from optimizing their distribution network for minimal cost, and towards maximizing value.

1.4.2   SUPPLY CHAIN DESIGN FOR A MULTI-NATIONAL PHARMACEUTICAL COMPANY

Problem setting. A multi-national manufacturer of pharmaceuticals operates a complex supply chain composed of various production and warehousing facilities. The company wishes to explore opportunities relevant to new network configurations (e.g., drop-shipping, integrated last-mile delivery) and modal shift (e.g., increasing the share of sea freight). They wish to differentiate supply-chain solutions with regards to the product family (e.g., adjust delivery frequencies and service levels depending on the product family margin). In addition, they wish to analyze the effects of potential network disruptions on the optimal network configurations. This Supply Chain Design problem is of *strategic* importance to the company, because of its significant long-term economic implications. The decision-making problem is *complex*, since product characteristics are widely different and there is no single Supply Chain Design strategy that optimally caters for all of them. The Supply Chain Design problem is also a *group decision-making* problem as it affects both the global supply chain strategy of the company and the local distribution approaches of individual country organizations, especially when it comes to serving certain key accounts. Therefore, to allow for an optimal design decision, it is critical to incorporate the tacit knowledge of local experts about the available infrastructure, country-specific last-mile distribution operations, and specific customer requirements.

   Application of visual analytics to support decision making. Similar to the previous case study, we implemented a visual analytics system consisting of a large-scale optimization model and an intuitively accessible interactive visual interface. The system was designed to serve as a *common decision-support tool* to be used by multiple user groups within the company. For example, it was intended to be used by strategic working groups on the global level tasked with identifying the right balance between sea and air transport of a variety of products, and in assessing the cost benefits of drop-shipping. At the same time, it was intended to support local country organizations in optimizing their replenishment strategy and last-mile delivery approach. Given the vastly different use cases the system was supposed to support, individual users needed to be able to *customize* the type of information displayed, and the visual representations used for this information. Every user group should be able to focus on the performance indicators and visual representations that were most meaningful for their specific use case.

Benefits of the approach. We can note multiple advantages of the proposed approach. On the process side, the proposed system allows more effective decision making and dramatically decreases the time required to assess various scenarios. In addition, it allows to simultaneously consider initiatives that were previously examined in a separate fashion by different working groups withing the company and to identify previously unforeseen synergies (e.g., links between drop-shipping and service-level adjustments). The proposed system has the ability to create a shared problem understanding through a common visual interface and to accommodate the individual perspectives of different decision makers through customized data visualization. This allows to bridge gaps between different teams and to enhance organization learning and cross-functional buy-in of different solutions proposed. In terms of business success, our approach generates significant cost improvements. Moreover, the ability to perform analysis on multiple levels and differentiate distribution approaches according to the product type and market region allows a move away from the company's historically grown one-size-fits-all approach. Alternative transportation modes, replenishment strategies, last-mile distribution approaches, delivery frequencies, etc. can now easily and reliably be evaluated for individual products and product families, allowing to avoid lost revenue potentials due to sub-optimal service-level propositions. Lastly, the visual analytics system allows to increase the resilience of the company's supply chain by assessing the impact of potential disruptions, supply shortages, or demand shocks.

1.5 CONCLUSION AND FUTURE RESEARCH

As supply chains have become a critical factor defining the competitiveness of companies, new approaches to Supply Chain Design are needed. In this chapter, we illustrate the combination of human-in-the-loop optimization and visual analytics as a way to enhance decision-making effectiveness and organizational buy-in. We draw from multiple research fields to establish a framework and a system design allowing to leverage complementary strengths of computers and expert human decision makers and to enable more comprehensive group decision-making strategies based on a combination of analytical reasoning and intuition.

The application to two case studies allowed us to demonstrate the benefits of such approaches in real-life organizational settings. Being able to adjust key model parameters through the visual interface and get immediate visual feedback on the effects of such a change on network performance allowed different groups to develop a shared understanding and to gain common intuition about the complex trade-offs they were facing. The "human-in-the-loop" was proven to be particularly effective in decision-making context where tacit business knowledge and information were distributed among multiple team members.

We can identify multiple avenues for future research in this area. First, implementing additional case studies would allow to further validate our framework and the proposed system design.

Subsequently, large-scale experiments based on our framework using synthesized data and case studies could be performed. On the one hand, such experimentation would allow to quantify the effects of the proposed system on decision-making effectiveness and solution acceptance. On the other hand, it would allow to investigate which strategies of human-to-human and human-to-machine collaboration yield best outcomes. Specifically, this type of experiment could allow to investigate the impact of the proposed system on cognitive biases in decision makers.

In addition, we encourage the academic community to further explore how variation in collaboration settings impact the decision-making effectiveness and outcomes. For example, future research could explore differences in team decision-making processes and decision-making outcomes between virtual and co-located teams and between synchronous and asynchronous modes of collaboration. Finally, future research could investigate further variations relevant to system design. For example, immersive environments have been shown to positively impact performance in data analysis tasks. Future research could focus on the role of such environments (e.g., larger spaces for manipulating data, ambient displays, multi-modal interfaces and the multi-channel communication) on team decision making.

REFERENCES

Alexander, E., S. Bresciani, and M. J. Eppler. 2015. "Knowledge scaffolding visualizations: A guiding framework." *Knowledge Management & E-Learning: An International Journal,* 7 (2): 179–198.

Andrienko, G., N. Andrienko, P. Jankowski, et al. 2007. "Geovisual analytics for spatial decision support: Setting the research agenda." *International Journal of Geographical Information Science,* 21 (8): 839–857.

Cacciabue, P. C. and E. Hollnagel. 1995. "Simulation of cognition: Applications." In *Expertise and Technology: Cognition and Human-Computer Cooperation,* edited by J.-M. Hoc, P. C. Cacciabue, and E. Hollnagel. Mahwah, NJ: Lawrence Erlbaum Associates, 55–73.

Calabretta, G., G. Gemser, and N. M. Wijnberg. 2017. "The interplay between intuition and rationality in strategic decision making: A paradox perspective." *Organization Studies,* 38 (3–4): 365–401.

Carter, C. R., L. Kaufmann, and A. Michel. 2007. "Behavioral supply management: A taxonomy of judgment and decision-making biases." *International Journal of Physical Distribution & Logistics Management,* 37 (8): 631–669.

Choi, T. Y., K. J. Dooley, and M. Rungtusanatham. 2001. "Supply networks and complex adaptive systems: Control versus emergence." *Journal of Operations Management,* 19 (3): 351–366.

Cilliers, P. and D. Spurrett. 1999. "Complexity and post-modernism: Understanding complex systems." *South African Journal of Philosophy,* 18 (2): 258–274.

Cohn, J., P. Squire, I. Estabrooke, et al. 2013. "Enhancing intuitive decision making through implicit learning." In *Foundations of Augmented Cognition: Lecture notes in computer science,* vol 8027, edited by D.D. Schmorrow and C.M. Fidopiastis. Berlin: Springer, 401–409.

Dane, E. and M. G. Pratt. 2007. "Exploring intuition and its role in managerial decision making." *Academy of Management Review,* 32 (1): 33–54.

Dong, Y., J. Xiao, H. Zhang, et al. 2016. "Managing consensus and weights in iterative multiple-attribute group decision making." *Applied Soft Computing,* 48: 80–90.

Evans, J. S. B. 2008. "Dual-processing accounts of reasoning, judgment, and social cognition." *Annual Review of Psychology,* 59: 255–278.

Fiore, S. M. and T. J. Wiltshire. 2016. "Technology as teammate: Examining the role of external cognition in support of team cognitive processes." *Frontiers in Psychology,* 7: 1531.

Franco, L. A. 2013. "Rethinking Soft OR interventions: Models as boundary objects." *European Journal of Operational Research,* 231 (3): 720–733.

Govindan, K., M. Fattahi, and E. Keyvanshokooh. 2017. "Supply chain network design under uncertainty: A comprehensive review and future research directions." *European Journal of Operational Research,* 263 (1): 108–141.

Heer, J. and M. Agrawala. 2008. "Design considerations for collaborative visual analytics." *Information Visualization* 7 (1): 49–62.

Hegger, D., M. Lamers, A. Van Zeijl-Rozema, et al. 2012. "Conceptualising joint knowledge production in regional climate change adaptation projects: Success conditions and levers for action." *Environmental Science & Policy,* 18: 52–65.

Innes, J. E. and D. E. Booher. 1999. "Consensus building and complex adaptive systems." *Journal of the American Planning Association,* 65 (4): 412–423.

Janjevic, M., D. Knoppen, and M. Winkenbach. 2019. "Integrated decision-making framework for urban freight logistics policy-making." *Transportation Research Part D: Transport and Environment,* 72: 333–357.

Julmi, C. 2019. "When rational decision-making becomes irrational: A critical assessment and re-conceptualization of intuition effectiveness." *Business Research,* 12 (1): 291–314.

Keim, D., J. Kohlhammer, G. Ellis, et al. 2010. *Mastering the Information Age: Solving problems with visual analytics.* Goslar, Germany: Eurographics Association.

Kirschner, P. A., S. J. Buckingham-Shum, and C. S. Carr. 2012. *Visualizing Argumentation: Software tools for collaborative and educational sense-making.* Berlin: Springer Science & Business Media.

Klein, D., H. Klein, and G. Klein. 2000. "Macrocognition: Linking cognitive psychology and cognitive ergonomics." *Proceedings from the 5th International Conference on Human Interaction with Complex Systems,* 173–177.

Klein, G., K. G. Ross, B. M. Moon, et al. 2003. "Macrocognition." *IEEE Intelligent Systems,* 18 (3): 81–85.

Klibi, W., A. Martel, and A. Guitouni. 2016. "The impact of operations anticipations on the quality of stochastic location-allocation models." *Omega,* 62: 19–33.

Kohlhammer, J. 2005. *Knowledge Representation for Decision-centered Visualization.* Berlin: GCA-Verlag.

Kohlhammer, J., T. May, and M. Hoffmann. 2009. "Visual analytics for the strategic decision making process." In *GeoSpatial Visual Analytics,* edited by R. D. Amicis, R. Stojanovic, and G. Conti. Dordrecht, Netherlands: Springer, 299–310.

Kozlowski, S. W. and G. T. Chao. 2012. *Macrocognition, Team Learning, and Team Knowledge: Origins, emergence, and measurement.* New York: Routledge/Taylor & Francis Group.

Lami, I. M. and L. A. Franco. 2016. "Exploring the affordances of collaborative problem-solving technologies in the development of European corridors." In *Integrated Spatial and Transport Infrastructure Development,* edited by H. Drewello and B. Scholl. Dordrecht, Netherlands: Springer, 65–80.

Le Pira, M., E. Marcucci, V. Gatta, et al. 2017. "Towards a decision-support procedure to foster stakeholder involvement and acceptability of urban freight transport policies." *European Transport Research Review,* 9 (4): 54.

Lebeau, P., C. Macharis, J. Van Mierlo, et al. 2018. "Improving policy support in city logistics: The contributions of a multi-actor multi-criteria analysis." *Case Studies on Transport Policy,* 6 (4): 554–563.

Lim, S. F. W. and M. Winkenbach. 2019. "Configuring the last-mile in business-to-consumer e-retailing." *California Management Review,* 61 (2): 132–154.

Liu, J., T. Dwyer, K. Marriott, et al. 2017. "Understanding the relationship between interactive optimisation and visual analytics in the context of prostate brachytherapy." *IEEE Transactions on Visualization and Computer Graphics,* 24 (1): 319–329.

Meignan, D., S. Knust, J.-M. Frayret, et al. 2015. "A review and taxonomy of interactive optimization methods in operations research." *ACM Transactions on Interactive Intelligent Systems (TiiS),* 5 (3): 1–43.

Melo, M. T., S. Nickel, and F. Saldanha-Da-Gama. 2009. "Facility location and supply chain management – A review." *European Journal of Operational Research,* 196 (2): 401–412.

Niu, L., J. Lu, and G. Zhang. 2009. *Cognition-driven Decision Support for Business Intelligence: Models, techniques, systems and applications.* Berlin: Springer, 4–5.

Park, S. 2017. "An exploratory study on the meaning of visual scaffolding in teaching and learning contexts." *Educational Technology International,* 18 (2): 215–247.

Pathak, S. D., J. M. Day, A. Nair, et al. 2007. "Complexity and adaptivity in supply networks: Building supply network theory using a complex adaptive systems perspective." *Decision Sciences,* 38 (4): 547–580.

Podolefsky, N. S., E. B. Moore, and K. K. Perkins. 2013. "Implicit Scaffolding in Interactive Simulations: Design strategies to support multiple educational goals." University of Colorado Boulder. https://arxiv.org/ftp/arxiv/papers/1306/1306.6544.pdf.

Rosen, M. 2010. "Collaborative Problem Solving: The role of team knowledge building processes and external representations." University of Central Florida. https://stars.library.ucf.edu/cgi/viewcontent.cgi?article=5213&context=etd.

Suthers, D. D. and C. D. Hundhausen. 2001. "Learning by constructing collaborative representations: An empirical comparison of three alternatives." European Conference on Computer-supported Collaborative Learning, 577–592.

Windhager, F., G. Schreder, and E. Mayr. 2019. "Designing for a Bigger Picture: Towards a macrosyntax for information visualizations." OSF Preprints. https://osf.io/q9kdt.

2 Resilience-based Analysis of Road Closures in Colombia
An Unsupervised Learning Approach

*Nicolas Clavijo-Buritica, Mastoor M. Abushaega,
Andrés D. González, Pedro Amorim, and
Andrés Polo*

2.1 INTRODUCTION

The analysis and evaluation of risks in economic systems is a constant today. Companies, communities, individuals, and governments are increasingly concerned about disruptive events that can affect their interests and well-being. Road infrastructure, for example, is a network that, if disrupted, could have significant impacts on supply chain actors and communities. Aligned with the above, this study aimed to analyze the behavior of the disruptions of the national (non-urban) roads of Colombia between 2018 and the beginning of 2019. Later, through a performance analysis, the impact of some types of disruptive events on productive sectors in Colombia is evaluated.

The analysis developed in this study is based on two main stages: the data mining stage and network flow analysis. The first stage was oriented to describe the risk levels and time to recovery (after disruptions) of the Colombian road network, from the application of unsupervised learning techniques and data visualization. In the second stage, based on a case study, this work allows us to relate and understand the potential effects that the disruptions could have in the flow of commodities of the main economic sectors along the logistic corridors that connect the main urban centers in Colombia. The results of this study, from a methodological and practical perspective, offer a starting point to support decision making in supply networks that seek economic benefits by anticipating risks. In addition, this study offers a methodology that can be used by government and private decision makers seeking to develop data-driven action plans for enhanced infrastructure and societal resilience.

DOI: 10.1201/9781003137993-3

21

2.1.1 PROBLEM STATEMENT

The logistic operation of all countries has a close relationship with the infrastructure that serves the different supply networks. For Colombia, more than 47.5% of its companies are in the interior of the country, far from the port areas and in most cases with at least one mountain range to cross to take the goods (or bring them to the interior of the territory) by road. Any closure of a national road can generate a disturbance, in the economy, society, and the environment. Consequently, this situation affects competitiveness due to logistics efficiency (Clavijo-Buritica et al. 2018). Hence, the Logistical Performance Index (LPI) considers infrastructure within its indicators, as any impact on infrastructure can have large repercussions on the logistics of a country (The World Bank 2018).

Since 2015, seven main logistic corridors have been formally defined for Colombia for cargo transport by road (Ministry of Transport 2015). See Figure 2.1 for all the logistic corridors. In Colombia, more than 80% of the cargo is transported by road, 16% by rail, and only 2% by river. In addition, Colombia has a score of 65 out of 100 on the Road Connectivity Index, below the average for Latin America (73) and the Organisation for Economic Co-operation and Development (84) (Consejo Privado

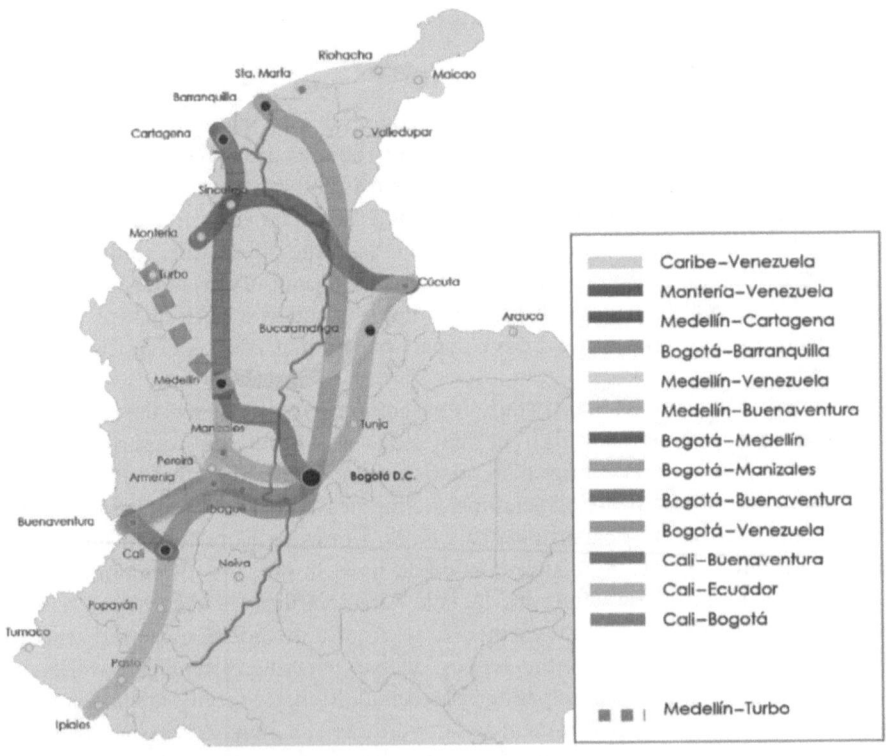

FIGURE 2.1 Colombian logistic corridors

Source: Adapted from DNP, 2008.

de Competitividad 2019). While the infrastructure of Colombia has improved in recent years, only 50% of the road network has been evaluated to determine its condition between paved or not paved (Valenzuela and Burke 2020). Additionally, Colombia not only has a large extension of flat territories but is also traversed by three mountain chains that are part of the Andes (reaching between 3740 and 5775 meters above sea level) which greatly increases the costs and complexity of road-based transportation.

The focus of this study is the analysis of road disruptions in Colombia due to total and partial closures. In the analysis, the main interest was in the *recovery time* (after disruption) and the *risk factor* of each event. Subsequently, the study focuses on understanding the incidence of these disruptions on the flow of products between major cities, since an analysis of un-met demand is closely related to resilience measures and provides information inputs for planning the restoration of logistic corridors (Zhao and Zhang 2020). Therefore, this chapter presents a practical and novel alternative to perform a data-driven analysis and then a network flow optimization. A multi-objective or stochastic optimization approach for resilience analysis is commonly used.

This chapter shows in the following five sections the results of the study conducted. Section 2.2 presents a literature review to answer: How has previous work studied resilience and disruption in road corridors?; and What data-driven and Machine-Learning-inspired methods have been used?. Section 2.3 provides a detailed description of the proposed methodology to simultaneously address uncertainties and infrastructure interdependencies in the road transportation network for a resilient distribution of commodities. Later, in Section 2.4, the main results are shown, illustrating the information pre-processing and corresponding modeling and generation of resilience metrics, in order to compare the behavior of the risk factors and recovery times for all regions in the country. In addition, the fourth section shows the association between road closure causes. Section 2.5 integrates the developed metrics with a network flows analysis, showcasing key dynamics of commodities between major consumption areas, along with potential implications. Finally, some conclusions and suggestions for future work are presented.

2.2   PREVIOUS RELATED WORKS

Resilience attends the four components of risk management: preparation, mitigation, response, and recovery (Fisher et al. 2010). A fundamental characteristic of resilient supply networks is to be adaptive, to anticipate risk, and to overcome disruptions (Reyes Levalle and Nof 2017). However, developing adequate resilience-based metrics and measurement methods for transportation infrastructure is particularly challenging due to their uncertainties and interdependences (Sun, Bocchini, and Davison 2020).

In recent years, some works have studied the disruption of roads and related infrastructure due to climate variability. For instance, traffic safety and congestion, associated with the disruption of road networks due to sudden changes in weather, especially highly variable rainfall in the mountainous areas of northern South America (Koetse and Rietveld 2009). These surveys reveal that although the study of road infrastructure resilience and disruption has increased in recent years, further research is needed in this field, particularly considering the increased frequency of disruptive events as a result of climate variables. In addition, through the study of vulnerability in transport systems, the authors have identified a lack of cross-disciplinary collaborations between policymakers, private sectors, and researchers to strengthen the resilience of transport systems (Mattsson and Jenelius 2015).

Under applications of statistical tools and data mining, there are several contributions, especially to analyze accidents, including the study of road traffic accidents in an Indian region (Atnafu and Kaur 2017), and the study of clearance time of roads after incident (Haule et al. 2019). The use of unsupervised learning such as clustering is common in traffic flow analysis and associated disruptions. For instance, the use of k-means to analyze the arterial traffic flow data (Gu et al. 2016), or clustering for classification of freeway traffic flow conditions (Azimi and Zhang 2010). From a risk assessment perspective, the risk evaluation for high-rise construction (Okolelova, Shibaeva, and Shalnev 2018) and the assessment of relations of risks and network topology (Zhang, Miller-Hooks, and Denny 2015) have been found. In addition, Bíl et al. (2015) analyzed the effects on the road network due to natural disasters.

Predictive applications have also been reported in the literature. The application of Support Vector Machine (SVM) and Artificial Neural Networks (ANN) to predict delays has been evidenced (Du et al. 2017). As well as the uses of Random Tree and Naïve Bayes to predict traffic incidents (Atnafu and Kaur 2017). While most previous studies related to the resilience of urban road corridors and associated infrastructure were oriented towards analyzing accidents, few contributions take a more holistic perspective on the disruption of non-urban road corridors. An exception is the study conducted by Muriel-Villegas et al. (2016) who presented an analysis of interrupted roads, and an approach where a complete statistical analysis allows to generate inferences on several variables, such as: vulnerability, reliability, capacity, and travel time, among others.

Given the scarcity of studies that do not deal exclusively with traffic accidents on urban roads or that emphasize disastrous events, it was our interest to delve into the research stream that addresses the study of resilience in regional (non-urban) logistic corridors. In this way, the main contribution of this work is the application of data analytics techniques in a real case of a country that has important challenges in terms of logistics and vulnerability related to disruptive events. Similarly, the methodology used and the management of the information to extract knowledge are a contribution to both researchers and practitioners.

2.3 SOLUTION APPROACH FOR RESILIENCE-BASED ANALYSIS OF ROAD CLOSURES

The methodological scheme designed to address the problem in this study was inspired by the phases of CRISP-DM (*CR*oss-*I*ndustry *S*tandard *P*rocess for *D*ata *M*ining) (Chapman et al. 2000). This scheme starts with an understanding of the problem and its associated data. It has a close relationship with the context in which the problem manifests itself (see Figure 2.2). This is to contemplate and understand that there may exist exogenous variables that can restrict the free behavior of some

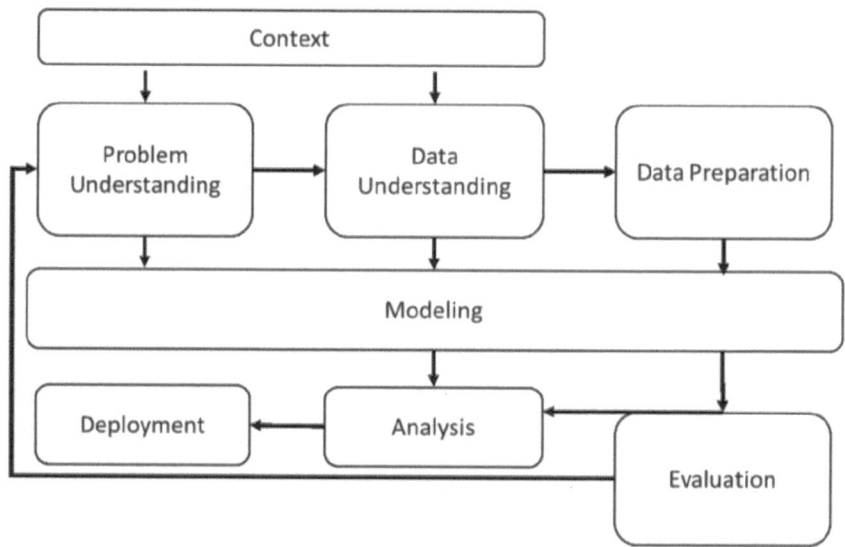

FIGURE 2.2 Proposed methodology inspired in CRISP-DM

internal variables. Then the data preparation, which is a critical process in the methodology, must be addressed.

When the data has passed through the preparation sub-tasks, the modeling should
be done, according to the need of the problem. Then, the applied models should be
evaluated using measures of accuracy, error, and other performance variables.

The *Context* of the problem was described in Section 2.1.1. The *Data Understanding*
in this study begins with an interpretation of the available information provided by
Colombian government entities. The information is semi-structured since it includes
a record of the date of each disruptive event, the department where it occurred, the
road, the type of disruption, and a long description (source of unstructured information about each road disruption). The information is composed of 29,127 disruption
events in 2018 and the first quarter of 2019.

The *Data Preparation* (pre-processing stage), was the most extensive and
important phase of the study. Although there were no duplicate events, there were
inconsistencies in the information that needed to be cleaned up. For example, a
column was generated to classify each interruption in total or partial road closure.
Some inferences were also generated based on logical rules in order to have more consistent information. In this last case, for instance, it was reviewed with conditionals
that two mutually exclusive cases were affirmed in two columns at the same time (i.e.,
case 1: total road closure, case 2: vehicular backflow). After debugging, combining,
and improving the information, two new characteristics were calculated for each of
the disruptive events. Since there was already a better and reliable classification of
the events, the risk factor and the estimated recovery time were calculated. This was
conducted using random values under certain inference rules based on expert knowledge. For instance, for the risk level of each disruptive event, a value between 0 and 6

is assigned, where 0 is no risk and 6 is extreme risk. Assignments are made according to the type of cause of the event. This ensures that an event associated with road maintenance work has a lower risk level than those closures generated by landslides or floods. Similarly, the assignment of estimated recovery days is carried out according to the cause of the road closure. For example, a closure due to heavy rain can take (for this study) a random value between 1 and 4 days, while a demolition would range between 6 and 9 days. These ranges depend largely on the agility of public and private institutions and their level of integration. All pre-processing tasks were carried out by using RapidMiner® and Excel®. In Section 2.4 of this chapter, more details are given about how the different pre-processing rules were used.

In the *Modeling* stage, although the pre-processing of the information generated structures that allowed the classification of disruptive events, both a description of the data and a clustering process were also performed. This was carried out to discover other interactions or relationships hidden in the data. Afterward, association rules were calculated (see Figure 2.3).

Both in the modeling and in the analysis, there was a primary interest in the recovery time after disruption, as well as the risk factor for each event. After the description of the data and its respective visualization, the clustering was modeled using the *k-means* model. In k-means (Zhexue 1998), given the group of elements X, we can find an assignment of each element to a specific group k considering a distance function. Its goal is to minimize the distances d between each X point and its centroid, as is presented below:

$$Minimize: P(W,Q) = \sum_{L}^{k} \sum_{i}^{n} w_{i,l} \, d(X_i, Q_1)$$

$$S.T. \qquad \sum_{L}^{k} w_{i,l} = 1, \quad 1 \le i \le n$$

$$w_{i,l} \in \{0,1\} \qquad 1 \le i \le n, 1 \le l \le k$$

In k-means the number of centroids can be selected in advance. Or this can be suggested by the software used. The latter case applies the well-known *X-means* where the best number of centroids in a range is selected for a dataset. For this study, each data point that must be assigned to a centroid in the clustering process is a road

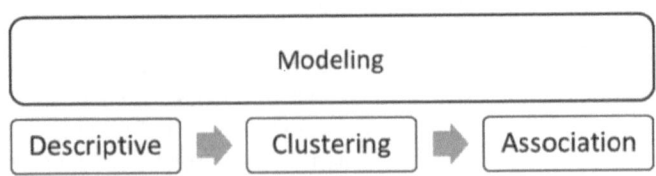

FIGURE 2.3   Modeling stage

closure (total or partial). In addition, since the number of centroids can be defined in advance (*apriori*) or can be optimized, for this study, the option of the optimal number of centroids is preferred.

For the development of the association rules between disruptive events, the *FP-growth* method, proposed by (Han, Pei, and Yin 2000), was used. The model is composed of two main phases, first the design and construction of the Frequent Tree (PT), which evaluates the frequency of certain items that are present in a set of transactions and which in turn meet a minimum threshold. FP-growth starts by eliminating items that are not frequent in the transactions, and then in a recursive process, obtains a reduced transaction database. In the context of this study, each disruption is a transaction of the original problem, while the type of disruption represents an item present in the transaction.

Once the unsupervised *Analysis* is developed, the methodology proposed in this chapter seeks to evaluate and relate the possible impacts that road closures had on the flow of products between major urban centers. In this way, by means of the well-known problem of flow networks with minimum cost (Bazara, Jarvis, and Sherali 2004), unsatisfied demand, product excesses, and later its relation with disruptions are analyzed.

The minimum cost flow problem was implemented to measure the flow of commodities in the Colombia road transportation network. The objective, Eq. 1, was to minimize Z, which represents the total transportation cost of delivering the required demand to the final demand nodes by minimizing the total un-met demand quantities and the excess supply quantities. The objective was subjected to the flow balance constraint, Eq. 2, and the road capacity constraint as formulated in Eq. 3, which considers the road functionality is not always 100%. The disruption level considered in this study was partial flow restriction through some of the networks' roads. After running the model, the supply chain network (SCN) performance was measured, based on the amount of the demand quantities that were not delivered to the demand nodes, then compared to the model performance when there were no disruptions.

Mathematical formulation:

Sets:

$\mathcal{N}$ set of nodes
$\mathcal{A}$ set of arcs
$\mathcal{L}$ set of commodities

Parameters:

b_{il} Net flow at node $i \in \mathcal{N}$ for commodity $l \in \mathcal{L}$
c_{ijl} Flow cost per mile of using arc $(i, j) \in \mathcal{A}$ for commodity $l \in \mathcal{L}$
u_{ij} The capacity of arc $(i, j) \in \mathcal{A}$

f_{il} Cost of un-met demand and oversupply at node $i \in \mathcal{N}$ for commodity $l \in \mathcal{L}$

y_{ij} A continuous parameter between $\{0,1\}$, 1 if arc is 100% operational; and 0 (zero) otherwise

Decision variables:

Δ_{il}^{-} Un-met demand at node $i \in \mathcal{N}$ for commodity $l \in \mathcal{L}$

Δ_{il}^{+} Excess supply at node $i \in \mathcal{N}$ for commodity $l \in \mathcal{L}$

x_{ijl} Amount of flow through arc $(i,j) \in \mathcal{A}$ for commodity $l \in \mathcal{L}$

Objective function and constraints:

$$Minimize \, Z = \sum_{l \in \mathcal{L}} \left(\sum_{(i,j) \in \mathcal{A}} c_{ijl} \, x_{ijl} + \sum_{i \in \mathcal{N}} f_{il} \left(\Delta_{il}^{-} + \Delta_{il}^{+} \right) \right) \tag{1}$$

$$\sum_{j:(i,j) \in \mathcal{A}} x_{ijl} - \sum_{j:(j,i) \in \mathcal{A}} x_{jil} = b_{il} + \Delta_{il}^{-} - \Delta_{il}^{+} \quad \forall i \in \mathcal{N}, \forall l \in \mathcal{L}, \tag{2}$$

$$\sum_{l \in \mathcal{L}} x_{ijl} \le u_{ij} \, y_{ijt} \qquad \forall (i,j) \in \mathcal{A}, \tag{3}$$

$$x_{ijl} \ge 0 \qquad \forall (i,j) \in \mathcal{A}, \forall \, l \in \mathcal{L}, \tag{4}$$

$$\Delta_{il}^{+} \ge 0 \qquad \forall i \in \mathcal{N}, \forall \, l \in \mathcal{L}, \tag{5}$$

$$\Delta_{il}^{-} \ge 0 \qquad \forall i \in \mathcal{N}, \forall l \in \mathcal{L}, \tag{6}$$

$$y_{ij} \ge 0 \qquad \forall (i,j) \in \mathcal{A} \tag{7}$$

$$y_{ij} \le 1 \qquad \forall (i,j) \in \mathcal{A} \tag{8}$$

The value of b_{il} represent the nature of the nodes, where:

- $b_{il} > 0$ node i is a supply node for commodity l
- $b_{il} < 0$ node i is a demand node for commodity l
- $b_{il} = 0$ node i is a transshipment node for commodity l

2.4 ROAD NETWORKS DISRUPTION ANALYSIS

The main results of the study are shown in the experimental phase. Initially, the pre-processing is shown, then the generalities of the description associated with the disruptions studied are presented. Finally, the results of clustering and association are reported. The data used includes the road disruptions in Colombia due to total and partial closures between the years 2018 and 2019.

2.4.1 PRE-PROCESSING

In the pre-processing phase, value-filling tasks were carried out (when there were missing values). The database had two columns that aimed to describe the disruptive events (State_1 and State_2). The first one had multiple errors, spelling errors, truncated values, and many types of declarations to say the same thing. The second one had only two possible values, "Restricted flow" or "Total closure". In order to have a consistent database, a simplification process was carried out. For badly written or misspelled records, the necessary combination and adjustment was made. In the same way for data that had several expressions to say the same thing, it was finally decided to leave only one or change to a better description.

On the other hand, by using software, the extraction of keywords from the column describing the causes of each road closure was carried out. This was done to generate more columns with values of "true" or "false" for the existence or not of a specific phenomenon. The new columns generated (as shown in Figure 2.4) provided more information for the subsequent analysis, and related to high-impact disruptive events for this research.

Given that the primary interest of the study is to analyze the behavior of disruptions on Colombian roads (non-urban), and their relationship with recovery times and the risk factor, the following illustrations are provided to visualize and profile the different regions of the country (see Figure 2.5). Additionally, all types of reported

FIGURE 2.4   Example for generating new columns (case: Landslides)

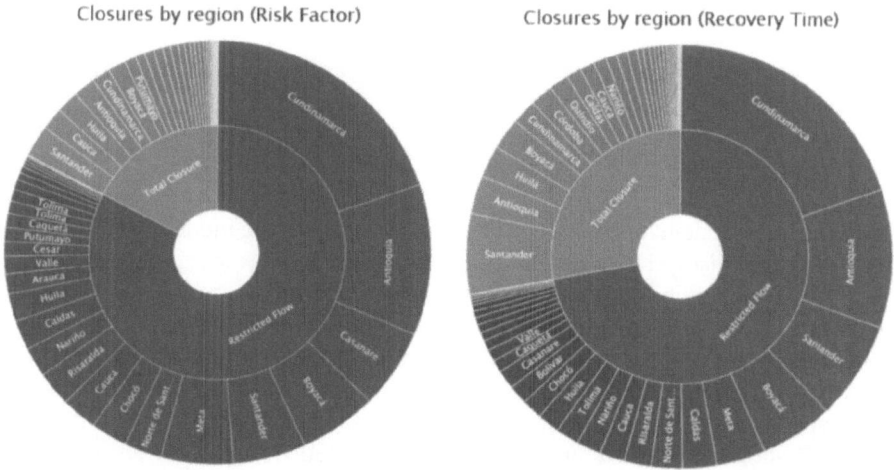

FIGURE 2.5   Closures by region: *(left)* risk factor in each region; and *(right)* recovery time in each region

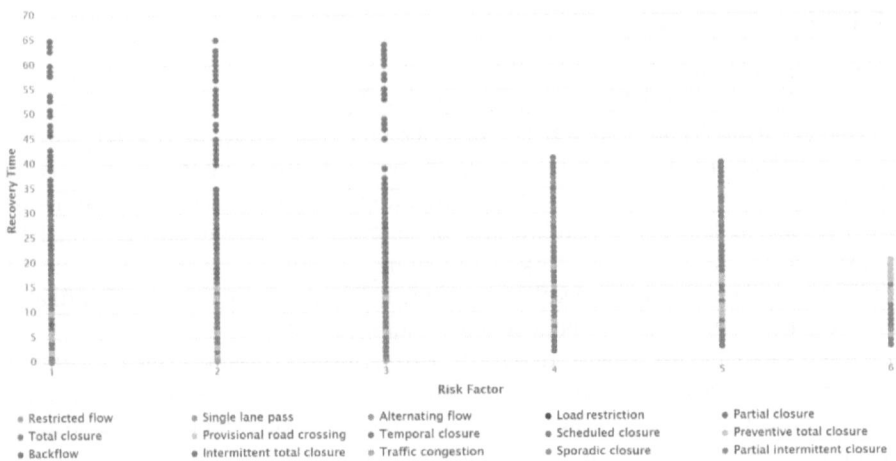

FIGURE 2.6 Relation between recovery time and risk factor for every reported disruption

Note: A color version of this figure can be found in the ebook.

TABLE 2.1
Evaluation of the Number of Clusters (k-means)

K	Score	Ranking
2	0.672	4
3	0.675	3
4	0.7	1
5	0.683	2

disruptions are shown and related to the risk factors and recovery times assigned for the 29,127 events (see Figure 2.6).

2.4.2 MODELING

The modeling was developed in two parts. First, the clustering model that allowed to unify sets for the disruptive events contemplating the risk and recovery time variables. The process was carried out using the k-means method and the number of centroids was selected from the score assigned to each one (see Table 2.1).

Since each point is assigned to a cluster, data visualization and analysis of the output report can now be used to identify which elements (disruptions) were most at risk or had the longest recovery times (see Figures 2.7 and 2.8). For instance, Figure 2.7 shows how the four clusters are distributed over the six levels of risk declared. Notably, clusters one and three tend to contain events with low and moderate levels of risk with high recovery times, while clusters two and four, although they have shorter recovery times, present high levels of risk.

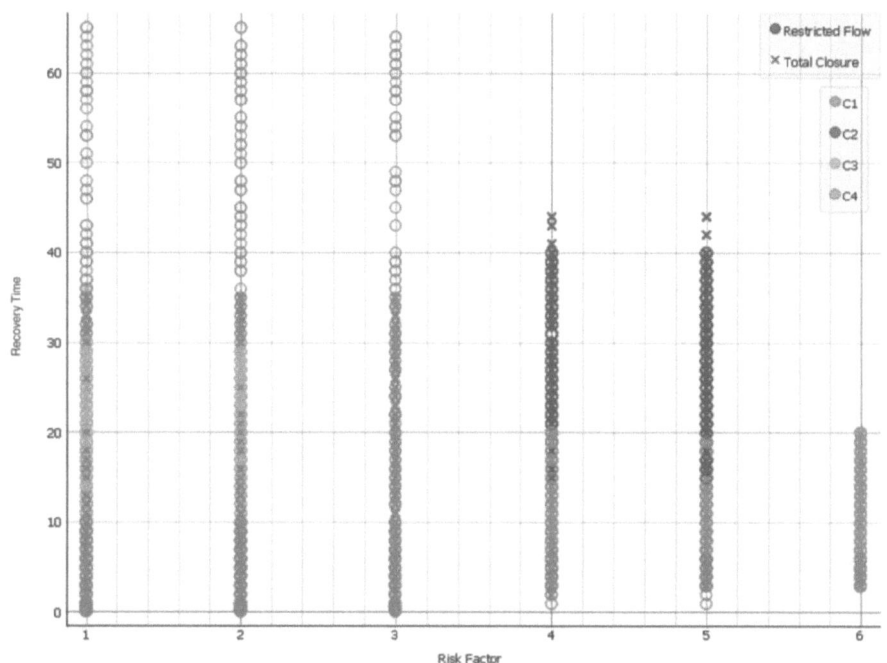

FIGURE 2.7   Recovery time and risk factor for four clusters

Next, for the application of the FP-growth model, two scenarios were carried out to find the rules of association. First, the target was to analyze the association of disruptions with total closures; and second, the association with partial closures. For the first case, the results are shown in Tables 2.2 and 2.3.

After the application of FP-growth, it is necessary that for the first case (target: total closures), the results are unreliable because of the small values of min support and min confidence. This result could be expected since the number of disruptions that generated total closures is less than the disruptive events associated with partial closures (restricted flow) and this algorithm works under the criteria of frequency of occurrence of the event. However, it should be noted that the technique allowed for the recognition that landslides play an important role and have a significant influence on total closures.

On the other hand, and with a higher level of confidence, more consistent rules of association can be found in the second case (target: restricted flow). The results of the second case are shown in Tables 2.4 and 2.5. For example, the high association of restricted flows with maintenance interruptions is evident. This is hardly logical. Similarly, a more interesting finding is found in the rule relating to restricted flows and landslides.

2.4.3   KEY FINDINGS

The clustering process generated relevant results, within which it was possible to analyze and identify that cluster No. 1 (C1) is the cluster that groups the events with the highest level of risk and with significant recovery times. In short, it is the cluster to

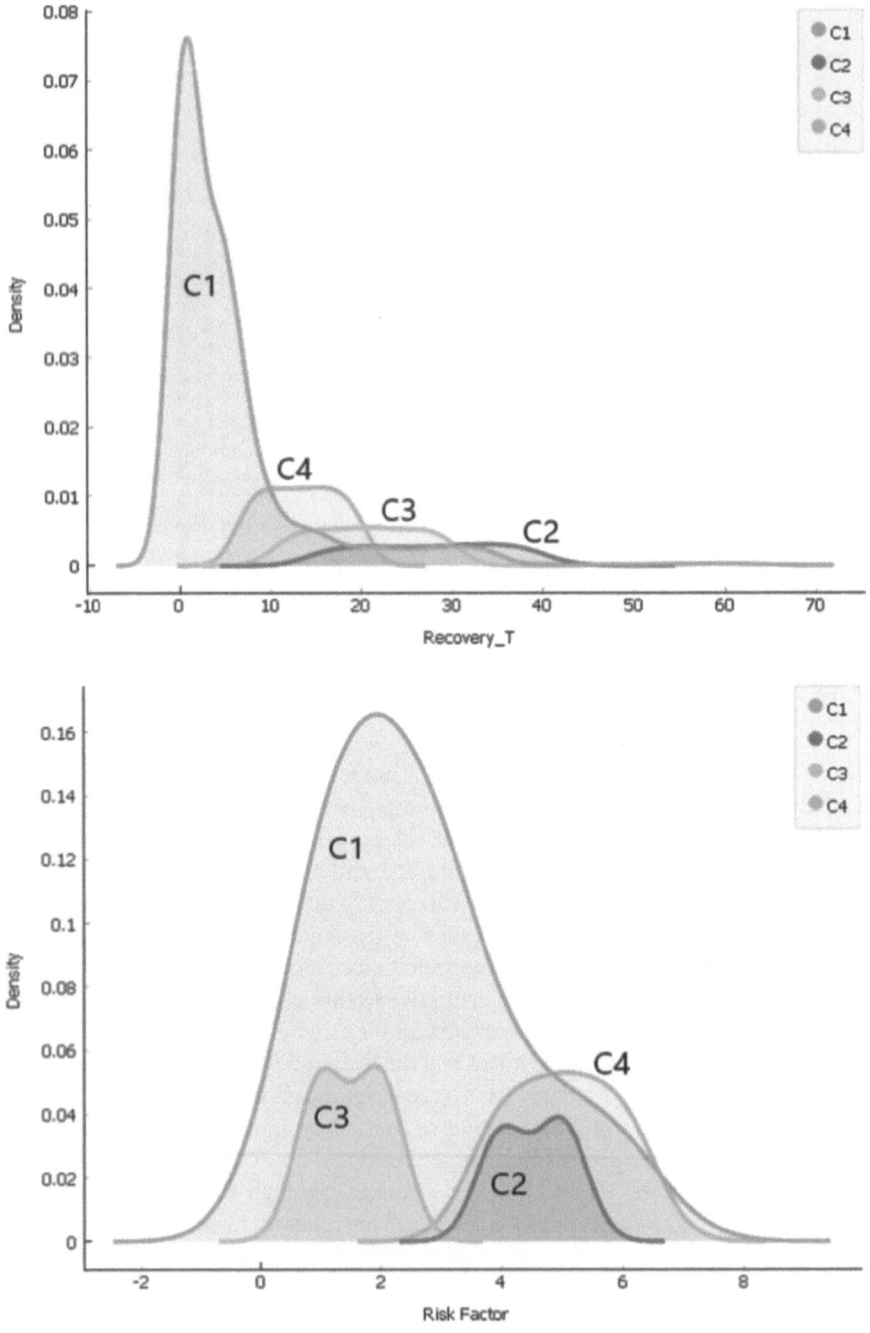

FIGURE 2.8   Distributions of recovery time *(left)* and risk factor *(right)* for four clusters

TABLE 2.2
FP-growth Results (target = total closures)

Support	Event 1	Event 2
0.203	Total Closure	
0.180	Landslide	
0.152	Road Maintenance	
0.126	Emergency	
0.074	Loss of Land	
0.019	Total Closure	Landslide
0.025	Total Closure	Road Maintenance
0.025	Total Closure	Emergency

Note: FP-growth support with min support = 0.09.

TABLE 2.3
FP-growth Results; Association Rules (target = total closures)

Premises	Conclusion	Support	Confidence	Conviction
Total Closure	Emergency	0.02499	0.1228	0.9964
Emergency	Total Closure	0.02499	0.1984	0.9936
Total Closure	Road Maintenance	0.02454	0.1206	0.9639
Road Maintenance	Total Closure	0.02454	0.1611	0.9495

Note: Association rules with min confidence = 0.1.

TABLE 2.4
FP-growth Results (target = restricted flow)

Support	Event 1	Event 2
0.797	Restricted Flow	
0.180	Landslide	
0.152	Road Maintenance	
0.126	Emergency	
0.161	Restricted Flow	Landslide
0.128	Restricted Flow	Road Maintenance
0.101	Restricted Flow	Emergency

Note: FP-growth support with min support = 0.4.

TABLE 2.5
FP-growth Results; Association Rules (target = restricted flow)

Premises	Conclusion	Support	Confidence	Conviction
Road Maintenance	Rest Flow	0.1277	0.8388	1.263
Landslide	Rest Flow	0.1610	0.8962	1.961

Note: Association rules with min confidence = 0.5.

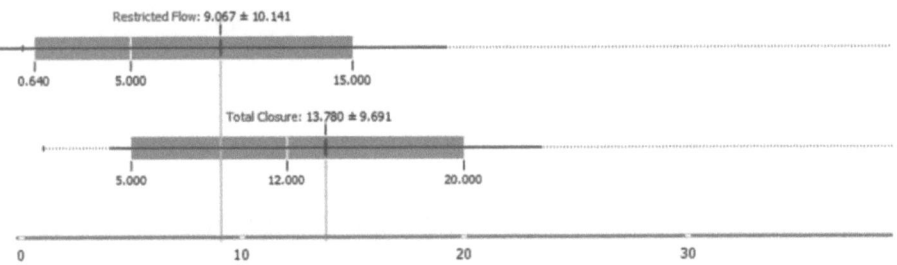

FIGURE 2.9   Box plot: recovery time (days) for total closures and restricted flow

which more attention should be paid. In addition, after the application of the association model in two scenarios, promising results were found only for the second where the target was to evaluate the association with restricted flows on interrupted roads. It is also necessary to mention that landslide interruptions have a high level of association with both types of events—total closure and restricted flow—but this was more obvious in the second case. Other descriptions could be made from the analysis of the graphed information, as well as from the same methodological route through which this study passed. Some of these are mentioned below.

In the information pre-processing phase, simultaneous descriptions were found for the same disruption. For those cases in which there was a temporary sporadic road closure and at the same time there was a road maintenance label, it could be inferred that it was corrective maintenance. In the same way, if there was a programmed road closure and a maintenance label was found in the description, it could be inferred that it was preventive maintenance.

Similar to the above, in pre-processing the information, cases were found in which the event was labeled as an inactive road and in the description, it was shown as a road maintenance. Here it could be inferred that it was a new road or a new bridge. From the box plots, it could be identified that the average recovery times after a disruption are different between total closures and restricted flow (see Figure 2.9)

It could also be identified from the box plots that the average risk factor regarding disruption type are different between total closures and restricted flow (see Figure 2.10).

Colombia has major challenges in terms of risk management and its relation to the sustainability of supply networks. A large expanse of the Colombian territory was found to be at risk of landslides, floods, and loss of road sections (as shown in Figure 2.11).

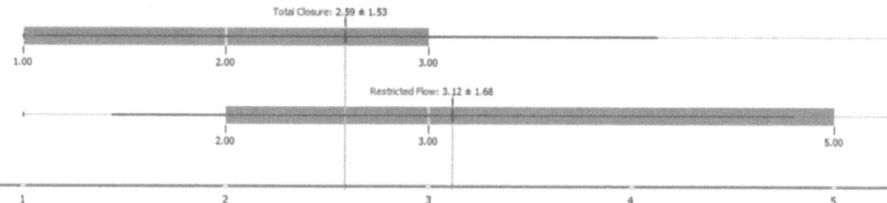

FIGURE 2.10 Box plot: risk factor for total closures and restricted flow

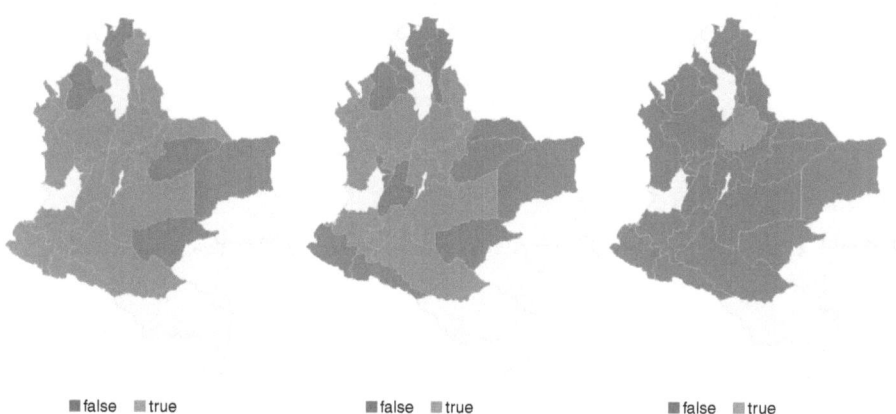

| ■false ■true | ■false ■true | ■false ■true |

FIGURE 2.11 Categorical map of Colombia and departments that had landslides and consequently road closures *(left)*; loss of road sections *(middle)*; and road flood *(right)*

2.5 EFFECTS OF ROAD DISRUPTIONS ON DOWNSTREAM SUPPLY CHAINS

In this section, the supply chain network (SCN) of several types of commodities is examined. The main transportation mode used in Colombia to distribute commodities is trucks. Thus, we can imagine the number of trucks transporting commodities in the road transportation network and the time it takes for each truck to reach its final destination. Figure 2.12 shows data for more than 250 products and reports the total amount of each product distributed in Colombia. Six commodities present around 33% of the total amount of products: Coal, Goods, Corn, Lime Cements and Plaster, Sugar, and Coffee. When the biggest Colombian exporter cities of these six products are examined in detail, it can be seen, in Figure 2.13, that Cartagena represents almost 10% of the total amount of the six products that are distributed in Colombia. Furthermore, five more cities and Cartagena represent 44% of the total exportation of the main six products in Colombia.

To measure the performance of the SCN, the 250+ products were classified into six sectors that represent six of the main economic sectors in Colombia (as presented in Table 2.6). 51 nodes (major cities in Colombia) and 57 roads connecting those cities were selected from the database expressed in this study. The number of flows entering

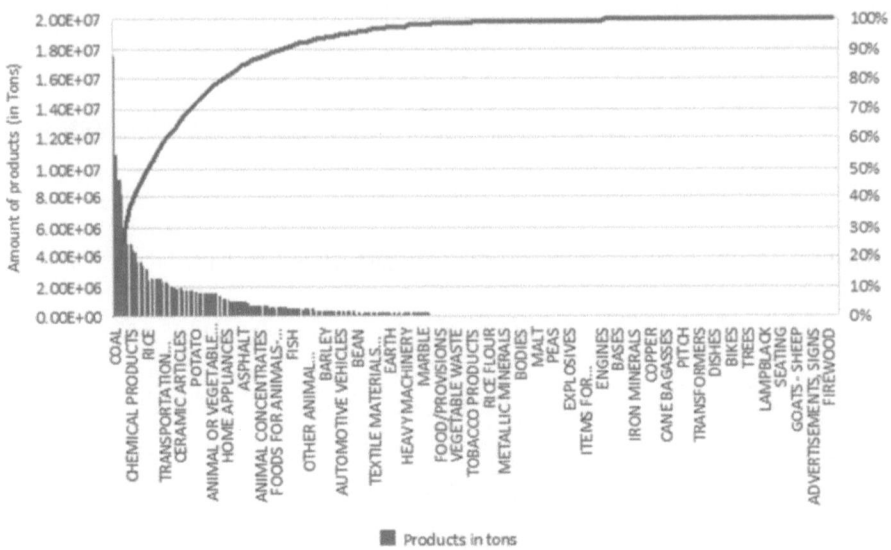

FIGURE 2.12   The total amount of different types of products distributed in Colombia

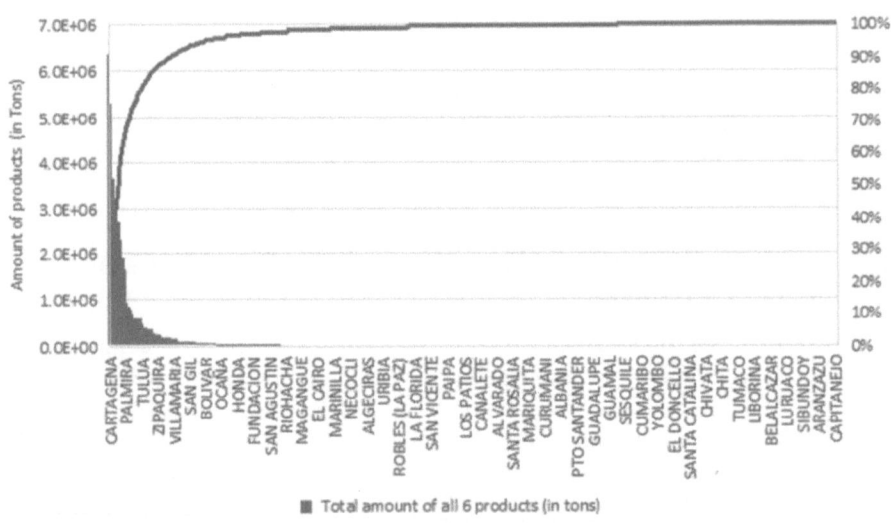

FIGURE 2.13   The big exporter cities of six of the main products in Colombia

and leaving the city (node) was considered in this case study to simplify the database. Thus, each sector has a set of supply nodes, demand nodes, and transshipment nodes, as depicted in Figure 2.14.

In Figure 2.14, each subfigure represents a specific sector with the associated node types. The link lines represent the flows of commodities in the network. The solid line represents the flow of material in the normal condition when there is no disruption

TABLE 2.6
The Economic Sectors in Colombia

Sector	Sector Description
1	Farming, forestry, hunting, and fishing
2	Mining and quarrying
3	Construction
4	Manufacturing industry
5	Commerce, repair, restaurants, and hotels
6	Transportation, storage, and communications

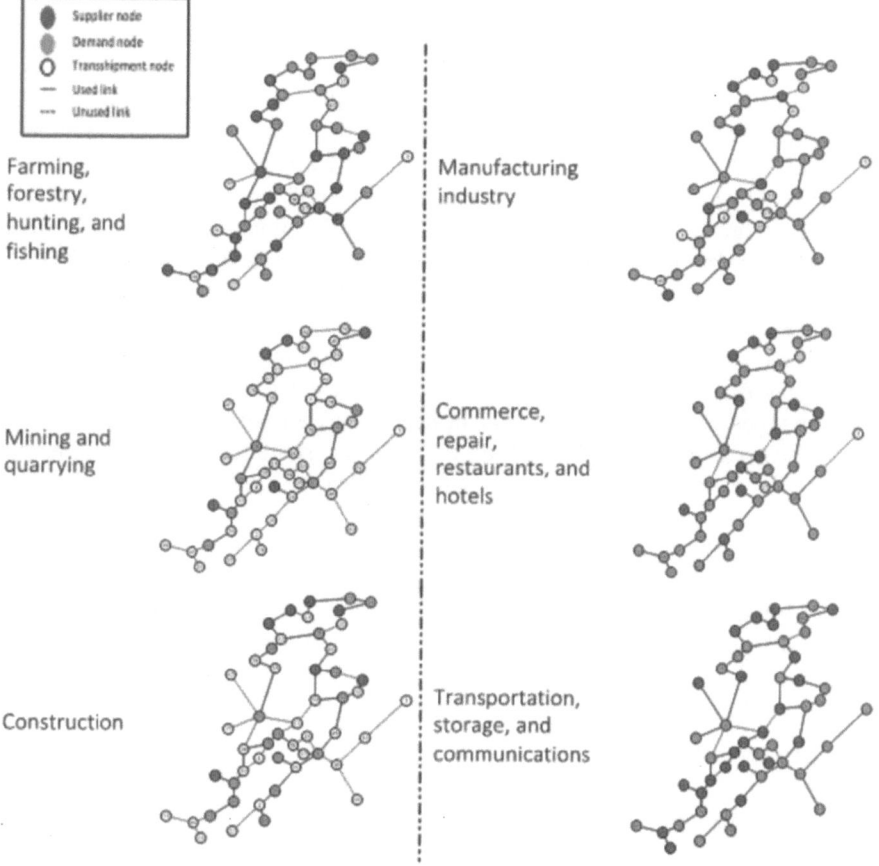

FIGURE 2.14 Node types of each economic sector in Colombia and the flow of commodities between nodes

disturbing the SCN performance. In contrast, the dashed line indicates that the link is not in use under normal conditions. The node shading defines the node types: dark (supplier nodes), medium (demand nodes), and clear (transshipment nodes).

As mentioned before, the primary transportation method to distribute products in Colombia is via trucks (Valenzuela and Burke 2020). Therefore, road functionality plays a significant and critical role in determining the efficiency of the SCN performance (KPMG 2019). Any disruption in the road transportation network prevents the flow from reaching demand nodes (cities). As a result, in this section, the impacts of disruptions on the road network are examined and measured. Due to Colombia's geographical properties, it is essential to study the effects of disruptions on the SCN performance.

The flow of commodities was examined under different types of disruptions that cause flows restrictions: landslides and road maintenance. 100 disruption scenarios were generated considering the obtained results in Section 2.4, in which the vulnerability of different regions for disruptions was presented (see Figure 2.11). To measure the impacts of the disruptions, the flow of commodities for all six main economic sectors was calculated in the normal condition when there were no disruptions in the system, and the system's performance when disruptions did occur. The results showed that the average performance of the 100 scenarios for each disruption type was impacted by more than 10%. The impacts of flow restriction caused by landslides were larger than the flow restriction caused by road maintenances. The SCN performance under the disturbance of each disruption event is depicted in Table 2.7. The landslides reduced the SCN performance by 20%, meaning that 20% of the total demand was not supplied to several cities. Moreover, the performance of the main six sectors was measured and demonstrated, as shown in Figure 2.15. Sectors 2, 3, and 4 (mining, construction, and manufacturing) were the most impacted sectors for both disruption types. Sector 1 (farming) performed the same for both disruption types.

From the given results, the impacts of disruptions in some countries where the road transportation networks are less redundant could lead to tremendous loss for other networks that rely on road networks distributing their products or providing their services. One of the main disruptions Colombia faces frequently is landslides that restrict road flows either partially or fully. In this section, we examined the impacts of the disruption on the flow material of several economic sectors in Colombia. The findings highlight the SCN capability to distribute the entire available product to the demand nodes (cities). But three sectors were the most impacted by disruption of landslides or flow restriction caused by road maintenances. Thus, it is recommended

TABLE 2.7
The SCN Performance During Disruption Events

Disruption Types	Avg. SCN Performance (%)	Std SCN Performance (%)	Confidence Level
Landslides	79.20	10.831	~ 95%
Maintenance	86.02	9.735	~ 95%

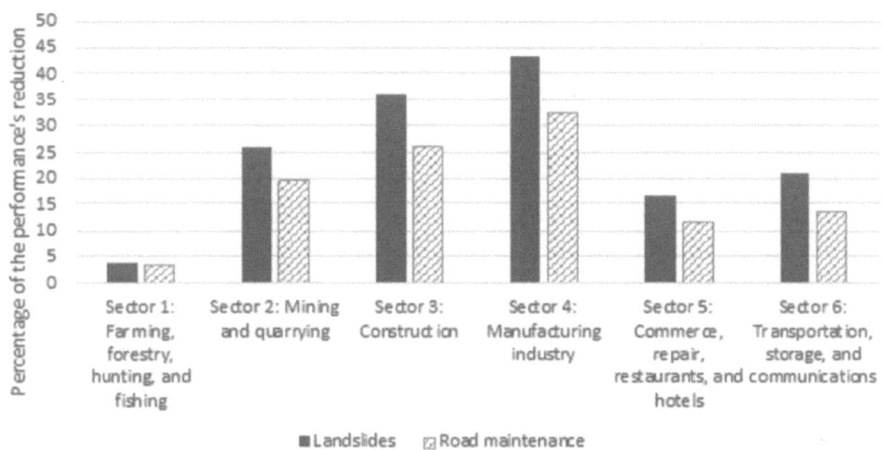

FIGURE 2.15 Comparison between the performance of each sector during different types of disruption

that some of the vulnerable areas be mitigated or retrofitted to enhance the road network durability against disruptions.

2.6 CONCLUSIONS

This chapter describes the disruptions on (non-urban) roads in Colombia between 2018 and the first quarter of 2019. From the application of clustering k-means and FP-growth, findings were made that may be useful for the management of the supply networks in the country. The methodological process was proposed from a slight adaptation of the well-known CRISP-DM, where the contextualization of the problem plays a key role in the modeling process, especially for the understanding and pre-processing of the data.

The pre-processing stage was the most important of all. It accounted for approximately 70% of the study and allowed more information to be generated from the processing of a difficult dataset. This was a column with 29,127 entries of descriptions of road closures. From this process, it was possible to extract new columns with data on the causes of closures.

From the obtained results, the vulnerable areas were determined and the impact in those areas were examined to measure the effects of such disruptions on Colombia's performance. The results showed how the total SCN performance was impacted and how each sector performed when the materials' flows are restricted due to landslides or road maintenances.

One of the main challenges of this study, apart from access to information, was the pre-processing phase because the volume of data was considerable. A major challenge for future work is the definition of risk levels for each of the disruptive events on the road network. In this work this task was developed through estimates and expert knowledge, however there are still significant research opportunities to develop methods and techniques for risk assessment of road infrastructure disruptions. Similarly, it is an opportunity to improve the accuracy of information on post-event recovery times.

ACKNOWLEDGEMENTS

Colombian Federation of Road Transporters—COLFECAR (Economic Studies) for providing the history of road closures and showing willingness to collaborate in the development of the research.

REFERENCES

Atnafu, Baye and Gagandeep Kaur. 2017. "Analysis and predict the nature of road traffic accident using data mining techniques in Maharashtra, India." *International Journal of Engineering Technology Science and Research,* 4 (10): 1153–1162.

Azimi, Mehdi and y Yunlong Zhang. 2010. "Categorizing freeway flow conditions by using clustering methods." *Journal of the Transportation Research Board,* 2173 (1): 105–114.

Bazara, Mokhtar S., John J. Jarvis, and Hanif Sherali. 2004. *Linear Programming and Network Flows.* Hoboken, NJ: Wiley.

Bíl, Michal, Rostislav Vodák, Jan Kubeček, et al. 2015. "Evaluating road network damage caused by natural disasters in the Czech Republic between 1997 and 2010." *Transportation Research Part A,* 80: 90–103.

Chapman, Pete, Janet Clinton, R. Kerber, et al. 2000. *CRISP-DM 1.0 Step-by-step Data Mining Guide.* The Netherlands: SPSS.

Clavijo-Buritica, Nicolas, Laura Triana Sanchez, Edgar Gutierrez, et al. 2018. "Resilient Supply Network Design: Disruptive Events Modelling for the Coffee Industry in Colombia." MIT SCALE Latin American Conference 2018, Boston, MA. www.researchgate.net/publication/349928252_Resilient_supply_network_design_disruptive_events_modelling_for_the_coffee_industry_in_Colombia.

Consejo Privado de Competitividad. 2019. *Informe Nacional de Competitividad 2019–2020.* Bogotá: Puntoaparte Bookvertising.

DNP. 2008. *Documento Conpes 3547 – Política Nacional Logística de Colombia.* Bogotá: Política Nacional.

Du, Bo, Steven Chien, Joyoung Lee, et al. 2017. "Predicting freeway work zone delays and costs with a hybrid machine-learning model." *Journal of Advanced Transportation,* 2017: 1–9.

Fisher, R. E., Gilbert Bassett, W. A. Buehring, et al. 2010. *Constructing a Resilience Index for the Enhanced Critical Infrastructure Protection Program.* Chicago, IL: Argonne National Laboratory.

Gu, Yiming, Zhen Qian, Xiao-Feng Xie, et al. 2016. "An unsupervised learning approach for analyzing traffic impacts under arterial road closures: Case study of East Liberty in Pittsburgh." *Journal of Transportation Engineering,* 142 (9).

Han, Jiawei, Jian Pei, and Yiwen Yin. 2000. "Mining frequent patterns without candidate generation." ACM SIGMOD Record, 29 (2): 1–12.

Haule, Henrick J., Thobias Sando, Richard Lentz, et al. 2019. "Evaluating the impact and clearance duration of freeway incidents." *International Journal of Transportation Science and Technology,* 8: 13–24.

Koetse, Mark J. and Piet Rietveld. 2009. "The impact of climate change and weather on transport: An overview of empirical findings." *Transportation Research Part D,* 14: 205–221.

KPMG. 2019. "The Road to Everywhere: The future of supply chain." https://assets.kpmg/content/dam/kpmg/xx/pdf/2019/11/future-of-supply-chain-the-road-to-everywhere.pdf.

Mattsson, Lars-Göran and Erik Jenelius. 2015. "Vulnerability and resilience of transport systems – A discussion of recent research." *Transportation Research Part A*, 81: 16–34.

Ministry of Transport. 2015. "Resolución No. 164." Bogotá, Febrero 5, 2015.

Muriel-Villegas, Juan E., Karla C. Alvarez-Uribe, Carmen E. Patiño-Rodríguez, et al. 2016. "Analysis of transportation networks subject to natural hazards – Insights from a Colombian case." *Reliability Engineering & System Safety*, 152: 151–165.

Okolelova, Ella, Marina Shibaeva, and Oleg Shalnev. 2018. "Development of innovative methods for risk assessment in high-rise construction based on clustering of risk factors." *E3S Web of Conferences*, 33.

Reyes Levalle, Rodrigo and Shimon Y. Nof. 2017. "Resilience in supply networks: Definition, dimensions, and levels." *Annual Reviews in Control*, 43: 224–236.

Sun, Wenjuan, Paolo Bocchini, and Brian D. Davison. 2020. "Resilience metrics and measurement methods for transportation infrastructure: The state of the art." *Sustainable and Resilient Infrastructure*, 5 (3): 168–199.

The World Bank. 2018. *Connecting to Complete, Trade Logistics in the Global Economy.* Washington DC: The World Bank.

Valenzuela, Lina S. and Michael J. Burke. 2020. "Toward a greater understanding of Colombian professional truck drivers' safety performance." *Transportation Research Part F*, 73: 188–204.

Zhang, X., E. Miller-Hooks, and K. Denny. 2015. "Assessing the role of network topology in transportation network resilience." *Journal of Transport Geography*, 46: 35–45.

Zhao, Tingting and Yu Zhang. 2020. "Transportation infrastructure restoration optimization considering mobility and accessibility in resilience measures." *Transportation Research Part C*, 117.

Zhexue, Huang. 1998. "Extensions to the k-means algorithm for clustering large data sets with categorical values." *Data Mining and Knowledge Discovery*, 2: 283–304.

3 Characterization of Freight Transportation in Colombia Using the National Registry of Cargo Dispatches (RNDC)

*Daniel Prato, Nury Rodriguez,
Juan Carlos Martínez, Camilo Sarmiento,
and Sebastian Talero*

3.1 INTRODUCTION

Road freight transport is the most crucial transportation mode in Latin America, accounting for more than 70% of the region's national cargo (Barbero and Guerrero 2017). Despite its relevance, the information generated on its performance is scarce, significantly less than that recorded for other modes of transport, such as rail, maritime, and air transport (Barbero and Guerrero 2017). In Colombia, the transport sector contributes 0.3% of GDP (Center for Economic Studies 2018). According to data from the Ministry of Transport (2017), 80% of the cargo movement in the country is carried out by road, so it is essential to facilitate its efficiency and productivity, providing agility and the necessary technology to improve the competitiveness of road freight (Triviño 2016).

Transportation is a diverse sector, ranging from small informal companies in the start-up phase to companies with a significant and constant demand for various sectors, such as urban distribution, waste collection, etc. All of them have a common problem: *the difficulty to collect and process data representative of activity or region in order to facilitate decision making to make the transportation process more efficient.* Currently, some institutions collect information on the sector from statistical entities; customs, tax agencies, and ministries, among others (see Table 3.1).

Currently, the information that is collected is scattered, and its analysis and integration are a challenge. The rise of open software platforms helps developers create

DOI: 10.1201/9781003137993-4

TABLE 3.1
Sources of Information on Land Freight Transportation

Types of Entities	Typical Information Collected
National statistical entities	Level of activity, freight rates, and cost structures
Automotive registrations	Vehicle registrations and characteristics
Business associations	Fleets and costs (some confidential information)
Vehicle manufacturers and marketers	Fleet sizes and sizes of fleets
Motor vehicle verification entities	Characteristics of verified vehicles
Tax agencies	Ownership structures and activity statements
Customs	International trades of motor freight transportation
Road safety entities	Road accidents and cargo vehicle involvement
Road entities, gauging, and origin-destination surveys	Transit, activity, and classification of trucks
Energy and environmental agencies	Emissions/consumption
Health agencies and insurance companies	Claims
Subnational governments	Traffic licenses
Private, confidential sources with relevant market data	Operational fleet utilization information (e.g., itineraries, congestion delays, delays in entering port nodes); freight rates
Consumer protection entities	Rates and prices of inputs

Source: Barbero and Guerrero, 2017.

applications for decision making by fleet managers and drivers. According to PwC's Global Transportation and Logistics CEO Survey (2018), based on 143 interviews with top executives worldwide, fleet management could be significantly improved through sensors inside trucks, ships, and aircrafts connected to Artificial Intelligence (AI) programs that monitor fuel consumption. They also recommend ways to minimize oil and gas use and programs that suggest predictive maintenance activities.

All of this information could be used to facilitate route planning, to design, optimize and manage deliveries according to customer schedule availability and/or delivery cost, as well as to aid vehicle maintenance, navigation traceability, and telematics-based integrated safety systems. There are also applications for sustainability purposes. The International Energy Agency (2015) developed the Mobility Model (MoMo) simulation to estimate and calibrate the energy consumption and GHG emissions of transport vehicles, based on data from 1990 to 2015. It considers vehicle sales, fleet sizes (stock), estimated fuel consumptions, mileage traveled, and truck occupancy/load factors.

Cardenas (2016) developed a prototyped system for decision support based on data-mining techniques for the behavior of logistic corridors in Colombia through the CRISP-DM 1.0 methodology and data from the National Registry of Cargo Dispatches (RNDC), analyzing about 180,000 observations. The analysis and processing of large amounts of information require the development of tools based on predictive analytics to identify patterns. These distribution functions represent a variable to predict over a time horizon.

Barbero and Uechi (2013) conducted an assessment of the availability and quality of transportation data in Latin America, analyzing information from eight countries (Argentina, Brazil, Colombia, Costa Rica, Mexico, Panama, Peru, and Paraguay). The main findings were:

- Regular coverage and acceptable quality for vehicle data.
- Low availability and intermediate quality for transport activity (task) data.
- Scarce availability and questionable quality for the transport of energy sources, and imperfect but reliable energy content by fuel type.
- Low availability and good quality for emissions data.
- Widely available and reliable economic and demographic data.

In Latin America, multiple initiatives have been carried out led by sub-regional organizations, universities, and national institutions, that seek to solve the lack of information that characterizes the region's logistics sector. The data reported by the observatories share indicators such as: kilometers traveled per year by type of vehicle; empty trips as a % of total trips; GHG emissions; accident rates; and types of cargo. However, these indicators vary from country to country, given the characteristics of each country. To standardize the indicators and generate a comparative analysis of performance in the region, the Inter-American Development Bank built a Logistics and Transportation observatory for 25 countries in Latin America and the Caribbean, presenting essential information through 53 indicators organized into six groups: general, roads, railroads, air, water, and logistics activities.

Regarding land freight transportation, it is evident that less information is recorded compared to the other modes. Most of the variables considered are *descriptive*. Even though countries such as Mexico and Spain have simulation tools in terms of transportation costs, and Chile has a cargo flow forecast, there is a gap in estimating future behavior that allows bold decision making for service optimization. In Colombia in particular, there are opportunities for improvement in indicators such as:

- Inclusion of economic variables, such as sector participation in GDP, employability, and sector costs associated with the operation.
- Crossover of road congestion variables versus delivery times.
- Forecasts of performance over the next three years in indicators such as emissions, loads, flow, and costs.
- Loading simulators for truck occupancy optimization.
- Route analysis maps, considering companies often have GPS in their fleets.
- Strategic points of cargo consolidation, which would make it possible to reduce the costs associated with transportation.
- A more synchronized method to update the data, which is challenging given the number of entities involved in the sector's indicators.

This chapter will address the results of the analysis of the information available in the RNDC of 2018 and 2019 to support the implementation of the National Logistics Policy (PNL) of Colombia by optimizing the processes of transportation, logistics, and national distribution of goods and merchandise. Moreover, connectivity with transportation networks and foreign trade nodes is promoted through an analytical tool to facilitate cargo transportation decisions.

3.2 METHODOLOGY

Freight transportation information in Colombia is concentrated in the logistics portal and is fed by the RNDC. It consolidates about 45% of the national-level trips, disaggregated by type of vehicle, type of cargo, freight, origin–destination, etc. The data comes from transportation companies and other agencies at the national level, such as the National Road Safety Agency and the National Roads Institute (INVIAS). Figure 3.1 illustrates the detail of the sources of input information that make up the control tower.

The next sections in this chapter follow the steps that were taken to process and analyze the information: data pre-processing, using a visualization tool, and identifying behavioral patterns.

3.2.1 DATA PRE-PROCESSING

Assessing the quality of information is critical, especially when there is a significant volume of data. For this study, we reviewed the information that road cargo carriers report to the RNDC by regulation in the cargo manifest, which, in general terms, consolidates the following data:

- The identification of the generator of the reporting load.
- Name of the freight transportation company that will provide the public freight transportation service.
- Description of the goods transported, indicating their weight or volume and origin.

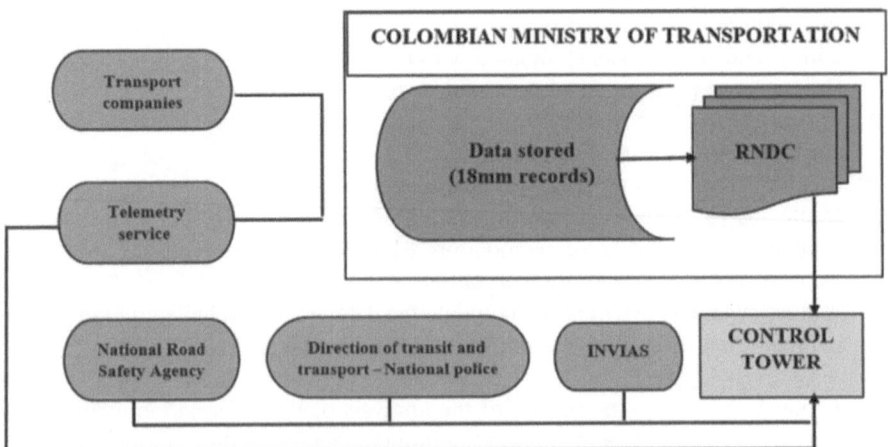

FIGURE 3.1   Sources of information that make up the control tower

TABLE 3.2
RNDC Data Analyzed by Year

Year	# Vehicles	# Trips	# Enterprises
2018	116,380	8.5 Million	2,067
2019	118,537	8.81 Million	2,053

- Place and address of origin and destination of the goods.
- Freight value.

Between 2018 and 2019, 17,575,821 records were generated (8,503,230 for 2018 and 9,072,591 for 2019). Table 3.2 presents a summary of the data analyzed by year.

As a first step for analyzing this information, a data cleaning process was carried out to identify atypical data, inconsistencies, differences, and similarities in the critical variables determined: tons, gallons, freight, kilometers traveled, total trips, and empty trips. The following criteria were used to clean the information that may contain inconsistencies:

- Manifest issuance dates as of January 2018.
- Financial variables must be greater than or equal to zero.
- Model year of the vehicle in operation after 1950.
- Values for which freight was less than 0 (Freight value: mean ± 2*standard deviation of data).
- Trips whose capacity is incongruent: From a master table of vehicles and their transport capacities (provided by the Ministry of Transportation), we search for vehicle fields whose recorded weight is less than the minimum according to the master table or greater than the maximum according to the same table (119 records).
- Trips with no kilometers recorded: Those trips whose origin–destination routes do not have a value related to kilometers (16,737 records).

3.2.2 Exploring the Potential of Data Through a Visualization Tool

Using Power BI (https://powerbi.microsoft.com/en-us/), analyzes were developed, with clean data, to understand the country's freight transportation information. Therefore, the information was systematically disaggregated by origin–destination, type of product, and vehicle type, according to the time distribution (weekly, monthly, semi-annual, and annual).

3.2.3 Identification of Behavioral Patterns in Freight Transportation

As a final step, we identified those travel behaviors by product type in a given season of the year that could present a repetitive pattern over time. We could then begin to work on predictions.

3.3   RESULTS

Understanding what the data reflects concerning cargo flow in the country is fundamental for the definition of actions and the identification of different actors' needs.
Therefore, this section describes the results obtained for each variable of interest.

3.3.1   PRE-PROCESSING OF INFORMATION

The value of freight reported per trip may depend on the type of product, distance
traveled, and other variables of consideration in the same negotiation. A comparison
of the raw vs. processed data for 2018 and 2019 is presented in Figures 3.2 and 3.3.

3.3.2   VISUALIZATION AND CHARACTERIZATION OF FREIGHT TRANSPORTATION

Python (www.python.org/) was used to visualize and statistically analyze the information from which the most used vehicles, the type of cargo, and their behavior over

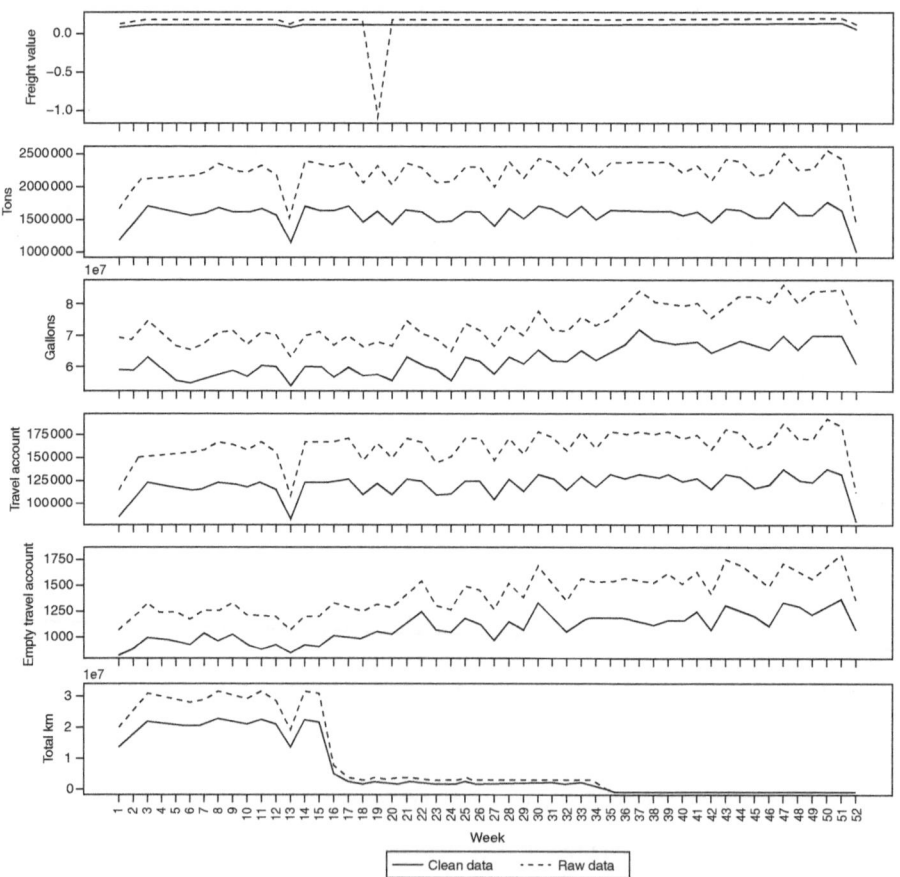

FIGURE 3.2   Cleaning of 2018 information

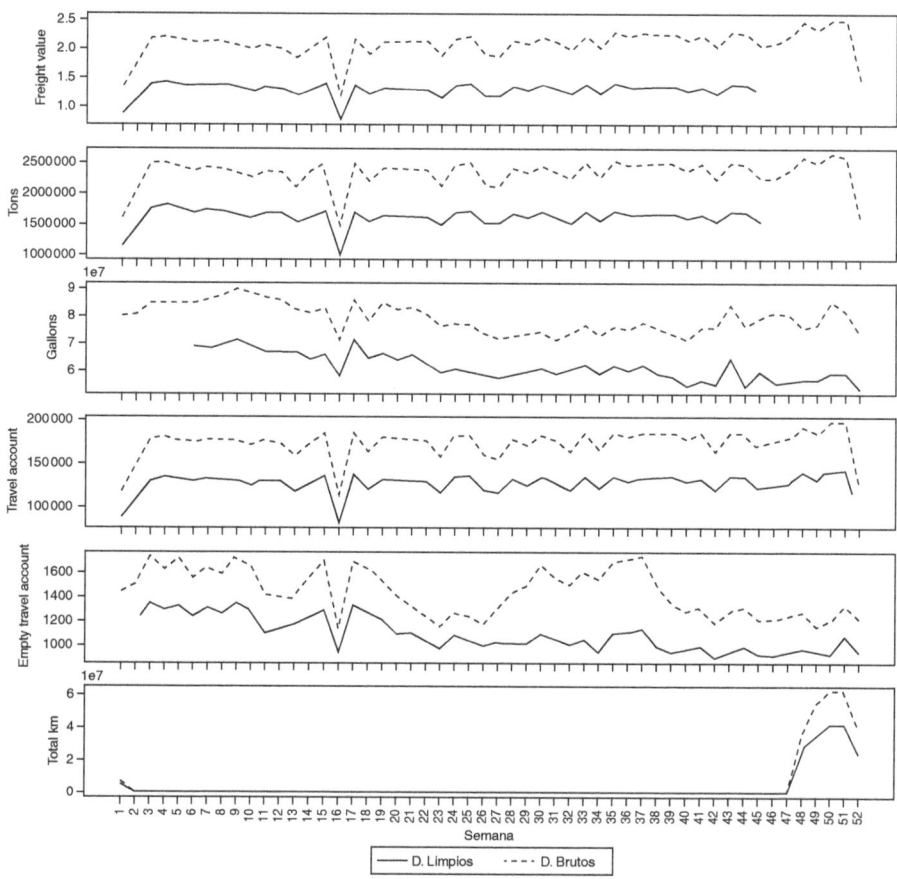

FIGURE 3.3 Cleaning of 2019 information

time were identified. A summary of the most relevant variables and their analysis and interpretation is presented below.

3.3.2.1 Types of Vehicles

For the period studied (2018–2019), it is observed in Figure 3.4 that approximately 50% of the cargo is mobilized mainly in type 2 trucks, which are commonly used for long-distance hauls and for large quantities of goods for nearby routes (see Tables 3.3 and 3.4). In addition, when comparing the maximum vehicle capacity and the cargo transported in Table 3.5, it is observed that vehicle capacity utilization is 63% for 3S2, which represents a potential to encourage actions to consolidate the country's cargo.

3.3.2.2 Main Origins and Destinations

Once the type of vehicles transiting the country's roads has been analyzed, it is useful to find out the most common origin and destination of transporters. This makes it easier to trace the routes in the movement of cargo at the country level during the

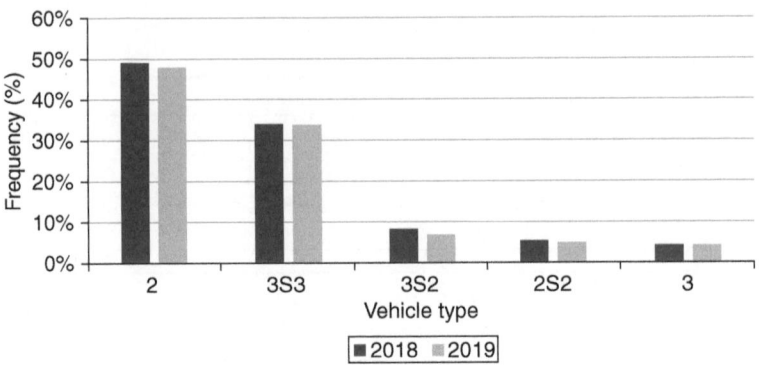

FIGURE 3.4   Distribution by type of truck

TABLE 3.3
Main Routes of Transported Cargo By Vehicle Configuration: 2018

Configuration	Route	Tons	Route	Gallons
2	Funza–Bogotá	274,113	Facatativá–Bogotá	3,058,118
	Yumbo–Cali	271,659	Puerto Asís–San Miguel	16,500
	Barranquilla–Cartagena	198,739	Yopal–Tauramena	7,600
3S3	Buenaventura–Guadalajara de Buga	1,588,689	Yopal–Tauramena	138,797,995
	Cucuta–Barranquilla	1,132,507	San Martin–Cartagena de Indias	124,258,584
	Buenaventura–Bogotá	1,124,579	Puerto Asís– San Miguel	111,923,507
3S2	Buenaventura–Yumbo	319,339	Yopal–Tauramena	12,800,119
	Buenaventura–Cali	190,932	Villanueva–Guaduas	41,483,357
	Cartagena–Bogotá	183,875	Facatativá–Bogotá	41,938,409
2S2	Buenaventura–Bogotá	337,299	Facatativá–Bogotá	6,532,987
	Cartagena–Bogotá	157,783	Yopal–Tauramena	9,689
	Buenaventura–Cali	77,146	Puerto Asís–San Miguel	540,489

TABLE 3.4
Main Routes of Transported Cargo By Vehicle Configuration: 2019

Configuration	Route	Tons	Route	Gallons
2	Yumbo–Cali	296,082	Facatativá–Bogotá	2,177,810
	Funza–Bogotá	319,288	Villavicencio–Puerto Gaitan	472,730
	Barranquilla–Cartagena	185,846	San Martin–Barrancabermeja	6,000
3S3	Buenaventura–Guadalajara de Buga	1,435,314	Yopal–Tauramena	127,138,463
	Buenaventura–Bogotá	1,195,166	Santa Marta–La Gloria	110,426,682
	Cúcuta–Barranquilla	1,134,580	Barrancabermeja–La Gloria	101,710,527
3S2	Buenaventura–Yumbo	374,975	Yopal–Tauramena	64,224,867
	Buenaventura–Cali	174,004	Villanueva–Guaduas	52,926,501
	Cartagena–Bogotá	191,371	Barranca de Upía–Monterrey	44,981,937
2S2	Buenaventura–Bogotá	309,474	Facatativá–Bogotá	1,531,274
	Cartagena–Bogotá	172,474	Barranca de Upía–Monterrey	840,994
	Buenaventura–Cali	76,030	Puerto Asís–San Miguel	845,065

TABLE 3.5
Vehicle Capacity Utilization

Description	Description Semi-Trailer	Configuration	% of Occupancy	% of Tons Transported	% of Gallons Transported
3-axle tractor-trailer	2 axle semi-trailer	3S2	63%	N/A	93%
	3 axle semi-trailer	3S3	56%	81%	96%
	semi-trailer with more than 4 axles	3S4	74%	91%	N/A

period of time studied, showing that in the period of 2018 Buenaventura was the destination that represented the largest cargo destination, while in 2019 Bogotá took first place (see Figures 3.5 and 3.6). This can be explained considering that the first is the main seaport of Colombia, and the second is the main economic center at national level (and so is the focus of trade with its large population). In terms of origin, the behavior is similar; both cities are positioned in the top two places for both years (see Figures 3.7 and 3.8).

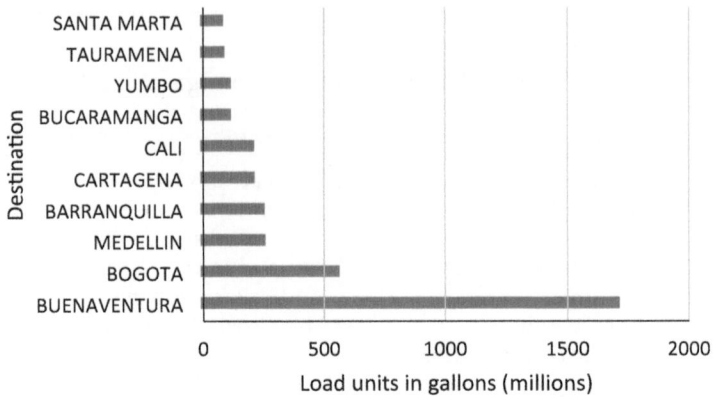

FIGURE 3.5 Top 10 cargo destinations in Colombia, 2018

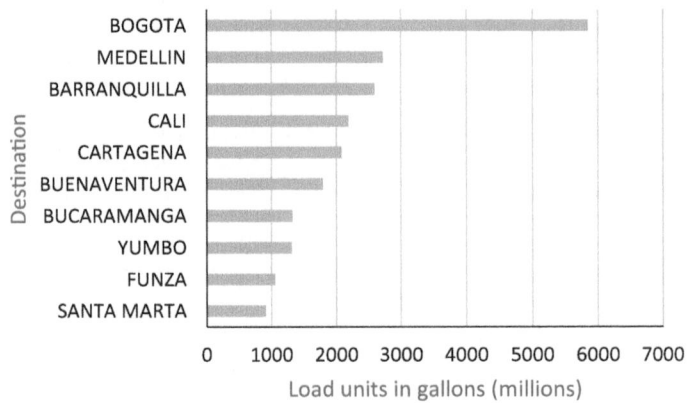

FIGURE 3.6 Top 10 cargo destinations in Colombia, 2019

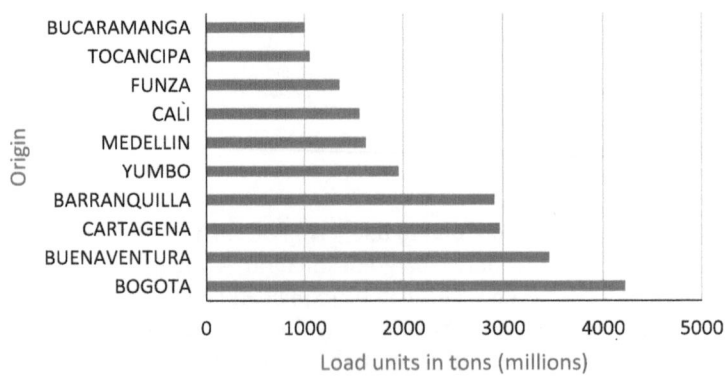

FIGURE 3.7 Top 10 cargo origins in Colombia, 2018

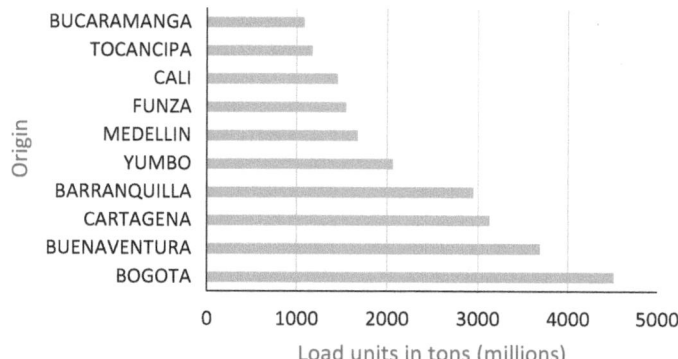

FIGURE 3.8 Top 10 cargo origins in Colombia, 2019

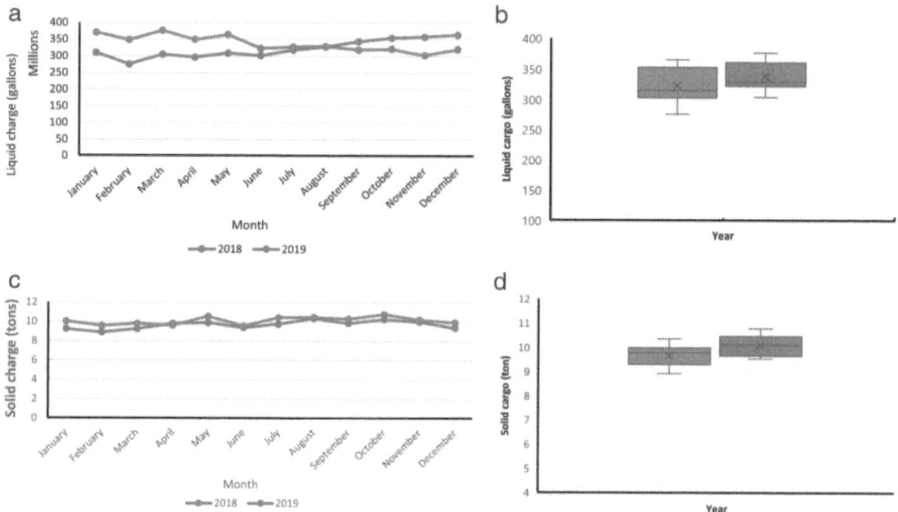

FIGURE 3.9 Cargo transported in 2018 and 2019: *(a and b)* liquid gallons; and *(c and d)* solid tons

From this point, it is possible to begin to disaggregate the information by type of cargo for 2018 and 2019. There are two major categories: liquid cargo and solid cargo. Initially, a characterization of the behavior by these two major categories will be made and then the detail of the main products for each category will be provided.

3.3.2.3 Liquid Cargo and Solid Cargo

The movement of both solid and liquid cargo for 2018 and 2019 had a similar behavior (see Figure 3.9). The troughs in 2019 for liquid cargo in the months of July and October is possibly due to the devaluation of oil-derived financial assets. Regarding solid cargo, in April 2019, there is a drop of 2.05% compared to 2018.

This is possibly due to the effects of production stoppage presented by some popular demonstrations, during which period restrictions in mobility and normal operation of some sectors of commerce began to happen, which generated a low demand for products. The box plots presented in Figures 3.9b and 3.9d show how the average transported cargo increased for the first half of 2019, with greater volume dispersion for certain categories.

Although the cargo movement in both categories (solid and liquid) for both years does not present significant changes, there is a difference in distribution patterns. For liquid cargo, there are two different patterns: in 2018, the cargo steadily increases in the second semester of the year; and in 2019, the cargo decreases in the second semester. In the case of solid cargo, the behavior patterns remain constant for both years.

3.3.2.4   Routes with the Highest Cargo Flow

Buenaventura–Bogotá transported ~3,424,636 tons of cargo in the last two years, and the most transported product on this route was cereal. Compared to 2018, this represents an increase of 9% in 2019. Other products transported include plastic materials, mainly on the Cartagena–Bogotá route, which during 2019 had an increase of 13.47% compared to the previous year. The Buenaventura–Buga route transports corn, but this decreased by 64.67% in 2019 compared to 2018. Other products such as densified wood and coal stand out on the Buenaventura–Cali and Cúcuta–Barranquilla routes respectively. Figure 3.10 illustrates the routes that transported the most tons in 2018–2019.

Liquid cargo presented variations in 2018 and 2019. During the first year the main routes were: Yopal–Tauramena, San Martin–Cartagena, and Facatativa–Bogotá. The latter had the biggest changes; the first quarter of 2018 presented growths of up to ~580% in March and decreases of ~560% for February and April. Together these routes transported ~814 million gallons. In 2019, the main routes were Cayaburo–Tauramena, Barrancabermeja–La Gloria, and Yopal–Tauramena, and they mobilized

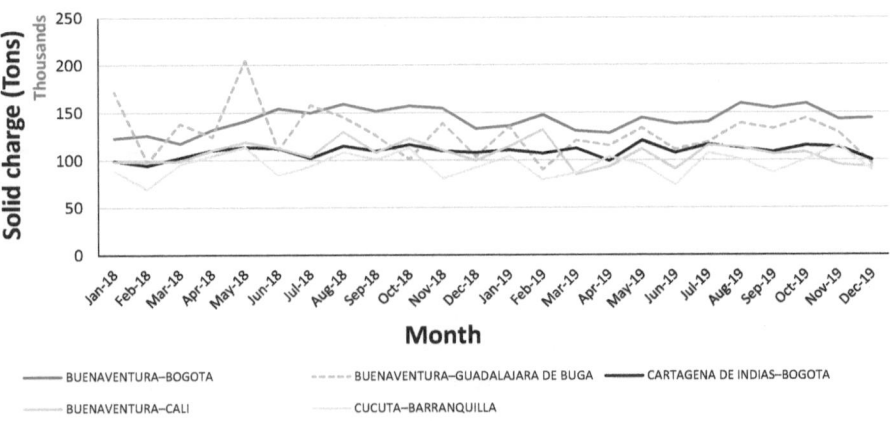

FIGURE 3.10   Routes that transported the most tons, 2018–2019

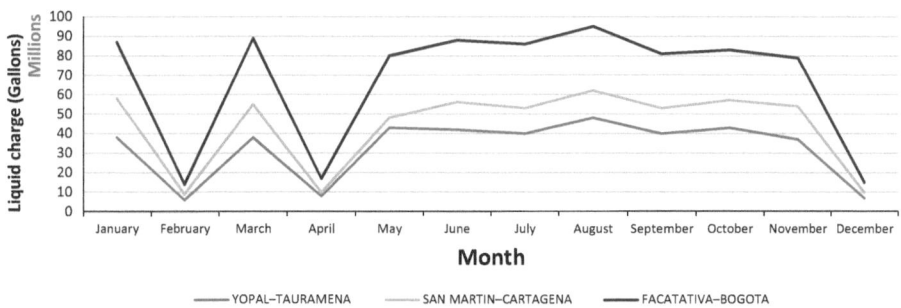

FIGURE 3.11 Routes that transported the most gallons, 2018

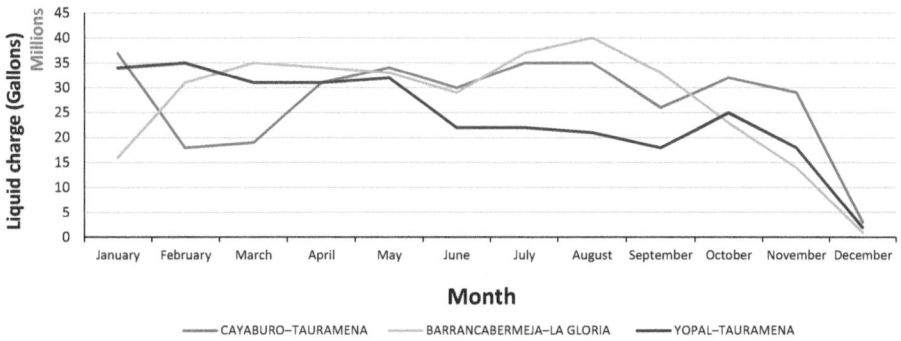

FIGURE 3.12 Routes that transported the most gallons, 2019

~946 million gallons. The last quarter of 2019 presented a 40% reduction in cargo compared to the first three quarters of the year. Figures 3.11 and 3.12 illustrate the routes that transported the most gallons in 2018 and 2019.

3.3.2.5 Most Transported Products

The most transported product in 2018 and 2019 was corn, registering a total of 5,946,931,615 tons. This is related to the increase in corn hectares in Colombia (increasing the production and supply) (see Figure 3.13). Regarding liquid cargo, the most transported product during the periods of analysis was crude oil at over ~3.1 billion gallons, followed by biofuels and fuel engine oil. The highest peak in the two periods was in July 2019, which had an increase for biofuels of 222% compared to the same month of the previous year. This is possibly due to the effect of the pricing policy, subsidies, and inflation (see Figure 3.14).

3.3.2.6 Variations in the Main Freight Transportation Variables

Although total cargo trips decreased by 0.3%, the number of vehicles increased by the same proportion. This can be explained considering that there was a 20% increase in gallons transported (see Figure 3.15).

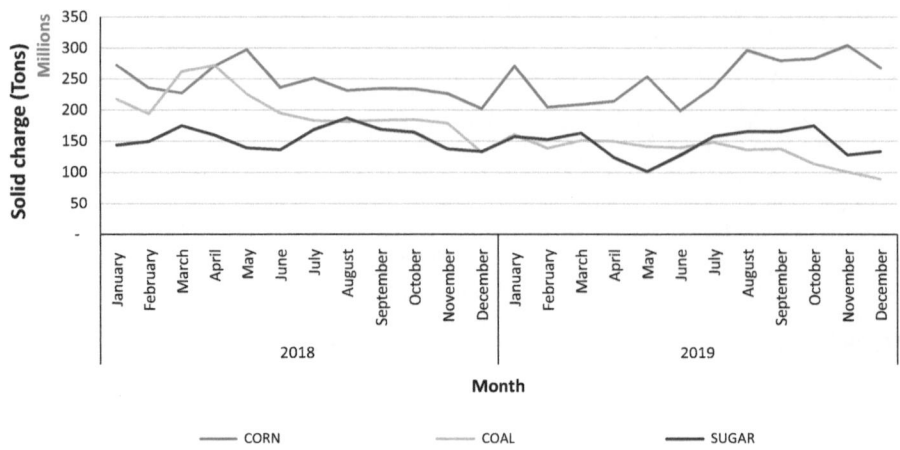

FIGURE 3.13   Top three most transported commodities in solid cargo, 2018–2019

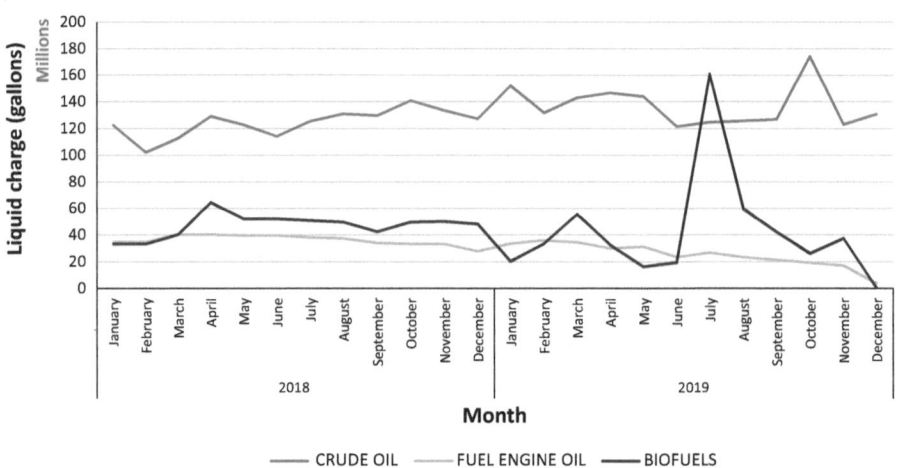

FIGURE 3.14   Top three most transported products in liquid cargo, 2018–2019

3.4 CONCLUSIONS

Freight transportation plays a fundamental role in the economic development of a country. Colombia has shown great progress in its logistics performance in the last decade, through the development of the logistics portal, which presents descriptive information on cargo movements considering factors such as: times (loading, unloading, route), types of vehicles (to determine cargo capacities depending on destinations), locations (in search of common cargo, determination of routes, vehicles), and routes (mainly, in terms of congestion, topology, stops), among others.

The constant data flows have demonstrated the need for sufficiently robust systems to store and facilitate their subsequent analytics. Annually, for example, the

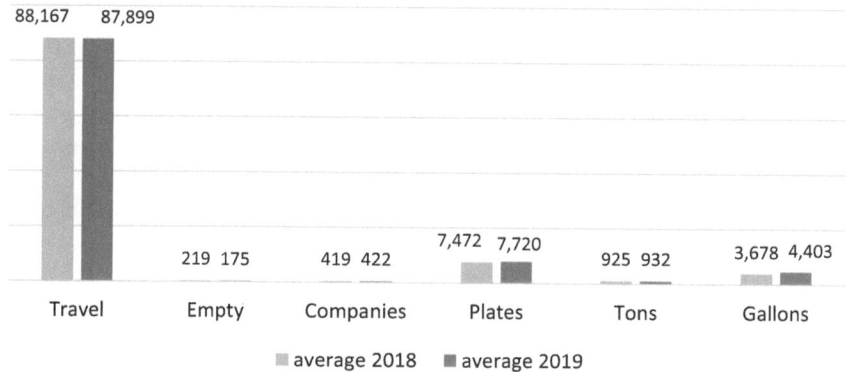

FIGURE 3.15   Main freight transportation variables in Colombia, 2018 and 2019

logistics portal received information from 17,575,821 records between 2018 and 2019 (8,503,230 for 2018 and 9,072,591 for 2019). The amount of data generated per day makes the processing of information one of the main challenges of the sector. This, coupled with the way in which the data is presented (descriptive in nature), restricts the possibility of making strategic decisions aimed at optimizing value network processes.

With the purpose of presenting alternatives to strengthen the dashboards of indicators available in Colombia's logistics portal on the national corridors, a data cleaning was carried out by identifying factors such as: trips whose capacity is incongruent (119 records), trips that do not record kilometers, minimum weight–maximum weight and minimum gallons–maximum gallons (in which outliers that did not correspond to the capacities considered in the SICETAC—Efficient Cost Information System for Automotive Freight Transportation—were eliminated). On the other hand, models older than 1950 were taken into account, and variables where data were mostly represented by zeros were omitted, such as kilometers (16,737 trips). Additionally, it was found that the most representative category in terms of trips was "other". This shows an opportunity for improvement in the methodology used for recording the information, to specify what should be included in this category, or whether the categorical options should be expanded, making it possible to analyze the behavior of specific products within the country. The probability of recording erroneous values increases considering that this is a manual process, as reflected in variables such as freight, where inconsistencies were identified with values that radically exceed the average value.

The result of the above exercise made it possible to define 21 variables for subsequent analysis, from which key aspects were identified, such as:

- Most goods were transported on a Friday during 2018 and 2019, accounting for 40,805,685 tons.
- The percentage of gallons transported increased by 20% in 2019 over 2018. However, the number of trips taken decreased by 0.3%.

- The percentage of vehicle occupancy increased slightly from 2019 to 2020, by 1.6% in solid cargo from 41.33% to 42.95% and in liquid cargo from 71.24% to 72.22%, which translates into an increase of approximately 1%. Therefore, it is essential to strengthen strategies to maximize the use of logistics resources in value networks: two of the best known are cargo compensation and consolidation.
- Bogotá is the main origin of cargo in the country, with 3,099,742.05 tons and 32,882,701 gallons going to different cities and municipalities.
- Empty trips are one of the main logistic challenges presented by the sector to increase efficiency in terms of occupancy. In this regard, it was evidenced that when considering the five routes that transported the most tons during 2018 and 2019, the percentage of empty vehicles decreased by 35%, despite the fact that the tons transported decreased by only 2%.
- The month with the most tons transported during 2018 and 2019 was August with a total of 10,359,271. In terms of gallons, December 2018 was the most productive month with a total of 364,534,258.

The review of available information, data cleaning, modeling, and dynamic diagramming of goods flows and logistic assets used by land transportation in the country allowed defining five ideas to strengthen the logistics portal. These were validated with different market profiles that make use of this tool, showing a trend mainly focused on the estimation of the logistic cost of transportation by product category.

For the development of the strategy, a methodology is proposed considering variables such as: freight, origins, destinations, categories with more trips, and weight configuration during the last years (2018, 2019, and 2020). Once the information is structured, it is proposed to consider the study conducted by LOGYCA (https://logyca.com/) on demand for the most representative product categories in the country. This study considers the same years for which information is available from the RNDC, which added to information on prices, origins, and weights, will allow having the visual of the variables considered in the logistics cost.

Finally, it is important to highlight that in order to continue strengthening this tool, the other four ideas should be explored: prediction of the number of trips that could be generated according to origin–destination and type of product; estimation of average waiting times to get a new trip depending on the load origin; and predictive simulation of traffic flow by corridor. This will add value to the sector's decision making. On the other hand, it is important to consider economic variables within the future analyzes that allow complementing the behavior of the sector such as: GDP, population, and employability, among others. Additionally, the possibility of establishing simulations should be explored.

REFERENCES

Barbero, J. A. and P. Guerrero. 2017. *Motor Freight Transport in Latin America: Logistical support for production and trade.* Washington DC: Inter-American Development Bank.

Barbero, J. A. and L. Uechi. 2013. *Assessment of Transport Data Availability and Quality in Latin America.* Washington DC: Inter-American Development Bank.

Cardenas, J. C. 2016. *Analysis and Prediction of Motor Freight Operation Behavior Using Data Mining Techniques*. Bogotá: Universidad Nacional de Colombia.

Center for Economic Studies (ANIF). 2018. *Transport Sector: Performance 2017 and Perspectives 2018*. Bogotá, Colombia.

International Energy Agency. 2015. "The IEA Mobility Model: A comprehensive transport modelling tool aimed at improving the analysis of all the aspects of mobility." www.iea.org/areas-of-work/programmes-and-partnerships/the-iea-mobility-model (accessed March 1, 2021).

Ministry of Transport. 2017. *Transport in Numbers*. Bogotá, Colombia.

PwC. 2018. "21st CEO Survey: A newfound confidence – Key findings from the transportation and logistics industry." www.pwc.com/gx/en/ceo-survey/2018/pwc-ceo-survey-transport-and-logistics.pdf (accessed March 1, 2021).

Triviño, Farfan. 2016. "Business strategies for the heavy cargo transport sector." Master Thesis. Universidad de Guayaquil, Ecuador.

4 Data and Its Implications in Engineering Analytics

Camilo Torres

4.1 DATA IS A VALUABLE RESOURCE IN ORGANIZATIONS

Data is the most important asset of companies today. It is considered the new oil of the twenty-first century (*The Economist* 2017). In the last two decades, organizations have increased their expertise in capturing, collecting, and storing data from their customers, operations, employees, and competitors. However, it is not enough to have data. We have to analyze this data, and this analysis can be a problem since organizations have to make essential investments in resources. Fifteen years ago, the technological components of storage had high costs, and it was not easy to procure them. Today, companies use interdisciplinary teams. These teams discover patterns, trends, and relationships. This process emulates the work performed in the last centuries where people were engaged in exploiting oil or minerals extraction. Today's miners have specialized tools (e.g., software, hardware, algorithms) that help them perform their data-mining tasks.

In 2018 the World Economic Forum (2018) warned of the GAFA + M concentration and its derivatives: **G**oogle, **A**pple, **F**acebook, **A**mazon, and **M**icrosoft. These companies have business models based on data. Netflix could also be added to this list with its streaming business model. Each of these companies has developed skills to compete through data analysis. In their way, each exploits the data of customers or users who use their systems or platforms. The World Economic Forum (2018) concluded that for every two dollars spent by customers of e-commerce in the U.S., Amazon is left with 50%. Amazon is considered the largest organization in the electronic trade. Amazon uses Big Data techniques that help to obtain valuable customer insight. The pandemic has benefited Amazon: millions of people in quarantine in different countries have turned to e-commerce to satisfy their needs. Amazon has doubled its profits in the second quarter of 2020 compared to the previous year, as shown in Figure 4.1. In September 2020, the price of Amazon's share on the NASDAQ was trading at $3,294. In December 2019, its price closed at $1,874; that is, in the last eight months, a growth of 76% (Financial statements for Amazon, Inc. 2021).

Google is the undisputed leader and dominant in online searches. Its famous search engine receives more than 4 million searches per minute, which generates millions of records that accumulate terabits and terabits (Big Data) of valuable information. This information is used for advertising purposes, among others.

DOI: 10.1201/9781003137993-5

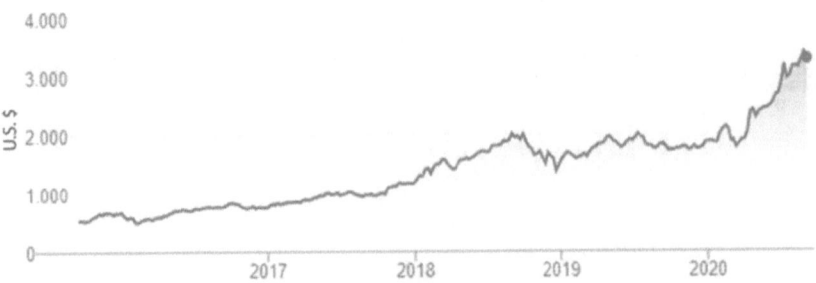

FIGURE 4.1 Amazon share price from September 2016 to September 2020

Source: Adapted from Financial statements for Amazon, Inc., 2021.

Another critical player in the career of the digital companies is Facebook. Facebook owns several companies, including Instagram and WhatsApp. With more than 2,449 million active users, Facebook undoubtedly dominates the global market for social networks. This number of users gives Facebook the absolute power to reach millions of people and influence purchase decisions. A British company, Cambridge Analytica, was involved in one of the most known data breach scandals in history. This company used Facebook to influence its users to manipulate voters (Chang 2018).

Microsoft has reacted with agility to the changes in the digital ecosystem. This reaction included the rethinking of its model, combining software with online services. Today, it is considered one of the most important companies in the world. Microsoft has made significant contributions in terms of data analysis. For example, Microsoft launched the first version of the spreadsheet (Excel) in 1985. Excel is used in the vast majority of organizations worldwide and is used daily by over 750 million people (Cocking 2017).

For its part, Apple Inc., an American multi-national company, specializing in the design and production of electronic equipment and software, is a leader in various market segments. Apple is positioned in the global market, with its iconic products iMac, iPod, iPad, and especially the iPhone. Apple is quoted on the market as worth a trillion of dollars.

Finally, we have the streaming giant Netflix, with a presence in more than 190 countries and more than 150 million active users. Netflix uses data from its users to improve their patented algorithm recommendations. These recommendations help them make suggestions to their users, and also aids in creating their series and movies. According to Todd Yellin, vice president of the company, Netflix knows at what time of the day the customers connect, how much time they spend on the platform, what they are watching and what they watched before, and even if the customers are using a computer, tablet, or mobile device. Their algorithm has evolved, and the company invests significant resources in its improvement. In 2006, Netflix launched the Netflix Prize (Netflix 2009). Netflix Prize is an open competition to improve collaborative filtering algorithms, which predicts the user ratings of movies based on previous qualifications and any other information about users or movies. Currently, the algorithm is so robust that it manages to ascertain users' preferences using Machine Learning.

TABLE 4.1
The Comparative Share Price of the Six Dominant Data-based American Companies

NASDAQ Companies	2015	2020	Variation %
Amazon (AMZN)	1,874.80	3,294.62	76%
Apple (AAPL)	74.36	120.96	63%
Netflix (NFLX)	329.09	516.05	57%
Facebook (FB)	208.10	282.73	36%
Microsoft (MSFT)	158.97	214.24	35%
Alphabet Inc (GOOGL)	1,354.64	1,581.21	17%

Sources: Financial statements for Amazon, Inc., 2021; Financial statements for Apple, Inc., 2021; Financial statements for Netflix, Inc., 2021; Financial statements for Facebook, Inc., 2021; Financial statements for Microsoft Corporation, 2021; Financial statements for Alphabet, Inc., 2021.

Table 4.1 presents the share price growth between 2015 and 2020 of the six afore-mentioned American companies. 2020 has been a good year for these companies, thanks to their models based on data and their technological capacity and adaptability to changes in the fourth industrial revolution—the digital era. The pandemic that paralyzed the world has strengthened and significantly improved the finances of these companies.

These companies are taking advantage of a massive data capture infrastructure (i.e., Big Data) and advanced analytics. They have achieved sustained growth that positions themselves as the world's largest companies. These organizations have found data to be their most important asset, so they invest significant resources in data capture, storage, and analysis. In addition, these organizations employ a talented scientific workforce (e.g., Geoffrey Hinton, who works for Google, is considered the father of Deep Learning: Hernandez, 2014).

4.2 A BRIEF HISTORY OF DATA ANALYSIS

Some authors think that data analytics techniques and Big Data are not new concepts (Borgman 2016; Raban and Gordon 2020). Initially, we disagreed. However, after a more in-depth analysis, we confirmed that several Artificial Intelligence (AI) techniques (e.g., neural networks) are not new concepts (i.e., just as concepts!). AI was introduced in 1956 at the Dartmouth Conference, where John McCarthy stated that "AI is a subdiscipline of the field of computing that seeks to create machines that can imitate intelligent behaviors" (Ertel 2011). The first model of a neural network was published in 1943 by researchers McCulloch and Pitts (1943) in their work: "A logical calculus of the ideas, immanent in nervous activity." However, those primary neural networks are very different from the current ones. Nowadays, the architectures are more sophisticated, and the algorithms use the latest developments of operations research and pattern recognition embedded in sophisticated computer systems. This type of advanced analytics is the basis of innumerable developments in AI. The boom of neural networks is accompanied by massive data and a significant increase in storage capacity and processing capabilities. Thanks to this, neural networks have facilitated the development of the visual computer area of the research field of capture, recognition, modeling, analysis, and generation of forms, images, and video.

Another example of early technique is the Turing test, formulated in 1950 by Alan Turing, considered one of the "computer science fathers." The Turing test is a test of a machine's ability to exhibit intelligent behavior equivalent to a human. Six decades later, for the first time, a machine using AI could pass the famous test. A chatbot (i.e., a robot programmed to chat online) convinced 33% of the judges who participated in the British Royal Society test that they were chatting with a 13-year-old Ukrainian boy (Veselov 2014).

To discuss descriptive analytics, we consider it pertinent to mention some significant milestones. Of course, they are just some examples without claiming that they are the only ones, but some of the most recognized.

Known as the lady of the lamp, Florence Nightingale (1820–1910), a British woman, left a vital legacy (Bradshaw 2020). Her postulate is currently the pillar of modern nursing, a profession created by Nightingale, who formalized training nurses at the beginning of the twentieth century. She became famous for treating the wounded and sick during the Crimean War (1854–1856). She, with a group of 38 nurses, found terrible conditions in a makeshift military hospital. She concluded that most deaths in this scenario were due to epidemic, endemic, and contagious diseases and not to wounds inflicted in battle. Her work in the military hospital in Turkey improved the living conditions of the soldiers. Thanks to her contributions, the military hospitals' mortality decreased significantly, thus obtaining the public's recognition in general.

At this point, the reader may be wondering what does nursing have to do with analytics? Florence Nightingale is also known for her outstanding contributions to statistics. Nightingale believed that statistics could be used to solve the problem of high hospital death rates. In the Crimean War, she began a rigorous process of collecting information with detailed daily records of admissions, injuries, illnesses, treatments to patients and its consequences, and soldiers' deaths. All these data, collected meticulously, was analyzed by Nightingale to understand the health problems and causes of death. She also designed a communication strategy of these findings, which materialized in one of the first examples of computer graphics—known as the diagram Nightingale Rose.

Nightingale compiled all her analyzes in her book *Notes on Matters Affecting Health, Efficiency, and Hospital Administration of the British Army*, of which few copies were printed, but one of which was sent to Queen Victoria of England. The innovative method of data analysis had a positive effect. Nightingale achieved her goal of revolutionizing and transforming nursing and health practices in the military. She became a personality and was appointed to the Royal Statistical Society in 1859, and the first woman to become an honorary member of the American Statistical Association in 1874.

Mary Eleanor Hunt Spear (1897–1986) worked as a graphic analyst at several U.S. Federal Government agencies between 1920 and 1960, including the Internal Revenue Service and the Bureau of Labor Statistics (Berinato 2019). She was the author of books *Charting Statistics* (1952) and *Practical Charting Techniques* (1969).

There are not many references to her work, so we refer to the research carried out by Ben Jones (2019), who describes her as a "pioneer in data visualization." In her first publication, a chapter entitled "The bar chart" shows an early form of the box plot that she called the "range bar," a graphical analysis tool that Tukey would later develop.

John Tukey (1915–2001) is an American statistician who left an important legacy in data analysis (Sande 2001). He got his Ph.D. in Mathematics from the University of Princeton and was director of the research group in statistics of this institution, where he spent most of his career. In 1965 he joined AT&T Bell Laboratories as the Chief of the Statistics Department. During this time, he participated in building electronic computers and developed the fast Fourier transform (FFT), which is essential for digital processing. John Tukey developed the Exploratory Data Analysis (EDA) approach. EDA is an essential contribution to what is currently known as data mining.

William Edwards Deming (1900–1993) is an American professor, statistician, consultant, and author who dedicated himself to spreading the concept of total quality management (Delavigne and Robertson 1994). One of his contributions is statistical quality control: the claim that the lower the variables in any process, the higher the quality of the resulting product. In 1921, he obtained a B.S. in Electrical Engineering, in 1925 a Master's degree in Physics and Mathematics from the University of Colorado, and in 1928 a Doctorate in Physics from Yale University, where he worked as a university professor. After the Second World War, in 1946, he visited Japan, where he was sent to study agricultural production and the war's adverse effects. In 1950, he was invited by several Japanese businessmen to give seminars on statistical control. These conferences were published and had great success, but it did not end there; his methods and theories began to be applied in many Japanese companies. Over time, managers' mindsets changed, and they began to control the quality of processes and materials. The use of tools and techniques of data analysis for statistical quality control spread in many countries from its influence on the productive Japanese sector. It also gave rise to the Six Sigma strategy improvement process that Bill Smith at Motorola developed. This approach is related to descriptive, predictive, and prescriptive analytics.

4.3   DESCRIPTIVE ANALYTICS

Davenport and Harris (2017) define descriptive analytics as business intelligence (BI) or performance reporting. Descriptive analytics provides the ability to alert, explore, and report using internal and external data from various sources. In their book, Davenport and Harris captured several studies and investigations in different organizations. These organizations have been developing their strategy in data analysis and analytical capacity: therefore, these companies are known as data driven. Such organizations have the following characteristics; they have a wide variety of quality data, integrated and available for use. These organizations have an analytical approach to management and informed decision making, using and analyzing data across the enterprise as an element of competitive differentiator. Another characteristic is that its highest level of management is committed to analytics, passionately

TABLE 4.2
Business Questions That Can Be Solved With Analytics

Analytics	Business Questions to Help Solve
Descriptive	• What happened? • How many, how often, where? • What exactly is the problem? • What actions are needed?
Predictive	• Why is this happening? • What if these trends continue? • What will happen next?
Prescriptive	• What happens if we try this? • What's the best that can happen?
Autonomous	• What can we learn from the data?

Source: Adapted and modified from Davenport and Harris, 2017.

promoting the use of data for decision making, thus aligning strategy and business objectives with analytics in such a way that different processes, metrics, and analysis of data creates a solid barrier that makes them effective analytical competitors. They employ experts who develop the necessary analyzes, using technology (software and hardware) to generate sophisticated analytics. Table 4.2 presents the four stages of analytics raised by Davenport and Harris and what business questions can be solved.

This chapter focuses on descriptive analytics, studying its nature, history, use in companies, and the available tools and benefits they bring to organizations. Analytics is a series of processes, tools, and techniques that help to understand the business. Businesses can analyze historical information and current trends and make decisions with greater strategic agility. With this process, valuable business knowledge is obtained for decision makers.

At the University of La Sabana in Colombia (www.unisabana.edu.co/), these data analysis techniques and processes are used to analyze millions of records. For example, data is taken from state tests to know how this institution is performing in each test. The results are compared over time to see their evolution, and they are contrasted between academic programs and between accredited universities that are references for this institution. They also analyze at a granular level, with students being the smallest grain available. This granular level allows other types of analysis to be carried out, combining statistical concepts that establish relationships with variables collected when entering the university and each student's academic development. Some academic success factors can be determined, understood as a good result in the state test that evaluates professional skills. Analytical reports relying on these data analysis tools are built.

The reports are shared with the different stakeholders such as government officials, deans, unit directors, program directors, professors, and students. This process helps to know precisely the strengths and weaknesses, and of course, to establish plans of action.

The analytical tools developed by the project team make it easier to identify patterns, trends, and relationships between the variables analyzed. All this helps to make informed decisions to validate the curriculum, identifying possible opportunities for improvement. The results, in turn, contribute to the academic reputation of higher education institutions. Another advantage of this type of analytical process and tools that this institution uses is identifying best practices and benchmarks.

4.4 VISUAL ANALYTICS

According to Davenport and Harris (2017), analytics is transforming decision-making processes in leading organizations worldwide. However, one of the great difficulties with data analysis is that it can be challenging to explain and understand. The general opinion is that analysts do not communicate well with decision makers and vice versa. Therefore, one of the final phases in any analytical process or project is the results' communication. The analysts must use tools that allow the message to be transmitted clearly, in such a way that anyone understands the findings found, so that decision makers can easily identify the necessary elements to guide informed decision making. Another skill that the analyst must have is Data Storytelling. Data Storytelling is the ability to tell stories with data to translate analyzes into simple terms to influence a decision. It is related to communicating efficiently and through visual elements to construct a narrative that facilitates understanding data, information, and sophisticated analysis.

Visual analytics is an entire area in data analysis, understood as a set of knowledge that allows us to use interactive visualization techniques with algorithms and data analysis methods to support analytical reasoning for decision making. Visual analytics provides a comprehensive, systematic framework for thinking about visualization in terms of design principles and options to facilitate understanding of the information. The graphical representation of information and data uses visual elements such as graphs and maps. Data visualization can detect and understand patterns, trends, relationships, and even anomalies, such as atypical values.

Visual analytics is applied in practically all fields of knowledge. In various disciplines, scientists use computer techniques and tools to model complex events and to visualize phenomena that cannot be observed directly, such as weather patterns and medical conditions. This visualization is vital to find relationships between variables or mathematical relationships such as financial assets' movements, among many others. In the current coronavirus pandemic, we find many examples of how different academic institutions, government entities, and national and municipal governments rely on analytical tools to track the statistics of the number of tests, infections, recovered cases, and deaths. An excellent example of visual analytics is Johns Hopkins University's coronavirus platform (Dong, Du, and Gardner 2020). This environment, developed by several groups at Johns Hopkins University, is one of the most widely used worldwide in order to monitor the pandemic's evolution. The World Health Organization (WHO) used the data to declare the new coronavirus outbreak of March 2020.

A critical aspect of data analysis is how the results are presented graphically. Making the insights more comfortable to understand depends mostly on the visual objects used to show the analyzes. When a report is presented, the human brain performs a series of processes; the pre-attentive attributes are processed in memory without conscious actions. In milliseconds the eyes and brain process these attributes in an image.

Ware (2019) defines four properties that facilitate pre-attentive processes and, therefore, understanding:

1. Color that includes saturation, hue, and luminosity. A carefully selected color palette helps harness the human brain's processing powers and makes information clearer, and more comfortable to find and understand. Colors can evoke a host of different emotions, from optimism, confidence, strength, and friendship, to defiance, fear, anxiety, and boredom. Colors can speak to the audience in many ways.
2. Attributes such as size, curves, length, marks, spatial clustering, and orientation.
3. Movement/animation can be used effectively to focus the attention of the public.
4. Spatial positioning is the quantitative attribute data encoded in the visual objects.

These elements are essential for visual analytics. With good management of these issues and attention to detail, data visualizations build a high impact that any audience can understand. Choosing the most suitable graphic is often a tedious task. The analyst must have the ability to find the visual object that helps to convey the message between the different categories of graphs: comparison, relation, composition, distribution, and geo-referencing. Complex graphics are good for exploring data but sometimes are difficult to understand. The goal is to tell a straightforward story using visually simple objects that the audience can understand, and then use this information and analysis to make decisions.

Consequently, classic charts like bar charts, line charts, and pie charts are better for communicating data. They make it easier to tell a story with the data. When telling stories with data, simplicity is vitally important. Well-designed dashboards, computer graphics, or presentations have a stronger impact and achieve the respective goals that people understand.

Although displaying data visually is not new, and these types of methodologies and techniques have been available for several decades, analysts often fall into the same errors and fail to convey their results. Thus the study and rigor of visual analytics is essential for professionals in the analytics area. It is a vital skill and the ideal complement to the technical knowledge in statistics, mathematics, and computer science that this type of professional requires.

4.5   ANALYTICAL TOOLS

The current market is saturated with analytical tools. The variety of technological resources and computational tools is so vast that companies and analysts can be overwhelmed when choosing the best option. Therefore, we will only reference some of the most common and well-known descriptive-analytical tools, without claiming that they are the only or the best. They are tools that facilitate information analysis processes and, therefore, decision making.

At a general level, there are two multi-purpose open-source tools: R statistical language and Python. R statistical language was developed at AT&T's Bell Laboratories (R Core Team 2020). It is one of the most commonly used programming languages by researchers and data scientists around the world. It was initially widely used for data mining, but its applicability has increased in other areas such as bioinformatics, mathematics, and Machine Learning. It is made up of modules that are easy to install. Each module comprises libraries containing the different functions necessary for the different data analysis types, particularly for graphical analysis. There is a library called GGPLOT. Thanks to this, the analyzes can be translated into visual objects that facilitate identifying patterns, trends, and relationships. Therefore, its potential is to facilitate the presentation of research results.

In 1991, Van Rossum (1995) created Python, considered by many to be the most powerful tool used by the data scientist community. It is a high-level object-oriented software with dynamic semantics, easy to use, with a smooth learning curve. It has a large number of specialized libraries for all types of data analysis. Some of the almost mandatory ones are *NumPy,* specialized for creating vectors and large multi-dimensional matrices and mathematical functions. *Pandas* is one extension NumPy uses for manipulation and analysis. *SciPy* is widely used in the engineering area since it has libraries for optimization, linear algebra, integration, interpolation, special functions, and signal and image processing. *Matplotlib* is a library to generate vectors and graphs from data contained in lists. It is one of the most commonly used for graphical analysis. These are the libraries that are almost always used in EDA, but they are just some of the libraries to be found in Python. Other libraries specialize in Machine Learning, and there is *Keras*, where all the modules for Deep Learning development are found.

For the above tools, specialized knowledge is required, especially in programming and statistical management. However, the following two tools are, according to Gartner (2020), leaders in the segment of analysis and business intelligence platforms:

1. From Microsoft, Power BI is a collection of software services, applications, and connectors that work together to turn unrelated data sources into visually compelling, interactive, and consistent information (Edge, Larson, and White 2018). Whether it is a simple Microsoft Excel workbook or a collection of local or cloud-based hybrid data stores, Power BI makes it easy to connect data sources, visualize or discover what is most important, and share it with whoever you want. The Data Analysis Expressions (DAX) language is used with Power BI. DAX is a library of functions and operators that can be combined to create formulas and expressions to potentiate the analysis. The combination of Power Query and DAX allows creating powerful reports with many data connected to different sources.

2. Tableau is a visual analytics platform that transforms how data is used to solve problems (Batt et al. 2020). The company was founded in 2003 as a result of a Stanford University computer science project. It has several products, specialized for different uses that help to integrate data, take it from different

TABLE 4.3
Top 10 Job Skills for 2020

1	Analytical thinking and innovation	6	Leadership and social influence
2	Active learning and learning strategies	7	Technology use, monitoring, and control
3	Complex problem solving	8	Technology design and programming
4	Critical thinking and analysis	9	Resilience, stress tolerance, and flexibility
5	Creativity, originality, and initiative	10	Reasoning, problem solving, and imagination

Source: World Economic Forum, 2020.

sources, consolidate it, transform it, graphically exploit it, and publish it. Along with Power BI, it is considered to be one of the best business intelligence (BI) tools. Tableau has tools that allow ETL (extract, transform, and load) processes to be carried out, making connections to different sources and origins, building dashboards that facilitate analysis, and telling stories with the data.

4.6   CONCLUSIONS

To finish this chapter, we would like to mention that, in the authors' opinion, the tool does not end the data analysis work. People who work in data analysis have different roles and inputs in the data enterprise, such as data analyst, BI expert, Big Data developer, data scientist, data engineer, analytics consultant, AI specialist, or chief analytics officer (CAO). We must take advantage of this technology to obtain insight, and to transform data into useful information to make informed decisions. We have to make the most of this valuable asset that, unlike other resources, does not diminish with its use. On the contrary, using data can build new data, and with the digital transformation in which most companies are immersed, we will continue to have much more data. It is important to be able to navigate without problems in this immensity of data and information.

The World Economic Forum (2020) published "The Future of Jobs Report 2020," a document that presents the skills most in demand for 2025. Table 4.3 lists the top 10 skills, and there are several related to data analysis.

The future of work and organizations will be significantly impacted by how collected data can generate a competitive advantage. With the support of different technologies and tools, analysts can make a difference by generating reports and data analysis models useful for decision makers.

REFERENCES

Batt, S., T. Grealis, O. Harmon, et al. 2020. "Learning tableau: A data visualization tool." *The Journal of Economic Education*, 51 (3–4): 317–328.

Berinato, S. 2019. *The Harvard Business Review Good Charts Collection: Tips, tools, and exercises for creating powerful data visualizations*. Boston, MA: Harvard Business Press.

Borgman, C. 2016. *Big Data, little data, no data: Scholarship in the networked world.* Cambridge, MA: MIT Press.

Bradshaw, N. 2020. "Florence Nightingale (1820–1910): An unexpected master of data." Patterns. www.cell.com/patterns/fulltext/S2666-3899(20)30041-6?_returnURL=https %3A%2F%2Flinkinghub.elsevier.com%2Fretrieve%2Fpii%2FS2666389920300416% 3Fshowall%3Dtrue.

Chang, A. 2018. "The Facebook and Cambridge Analytica scandal, explained with a simple diagram." Vox. www.vox.com/policy-and-politics/2018/3/23/17151916/facebook-cambridge-analytica-trump-diagram.

Cocking, S. 2017. "Seven Reasons Why Excel is Still Used by Half a Billion People Worldwide." Irish Tech News. https://irishtechnews.ie/seven-reasons-why-excel-is-still-used-by-half-a-billion-people-worldwide/ (accessed March 1, 2021).

Davenport, T. and J. Harris. 2017. *Competing on Analytics: The new science of winning.* Boston, MA: Harvard Business School Press.

Delavigne, K. and J. Robertson. 1994. *Deming's Profound Changes: When will the sleeping giant awaken?* New York: Pearson Education.

Dong, E., H. Du, and L. Gardner. 2020. "An interactive web-based dashboard to track COVID-19 in real time." *The Lancet Correspondence,* 20 (5): 533–534.

Edge, D., J. Larson, and C. White. 2018. "Bringing AI to BI: Enabling visual analytics of unstructured data in a modern business intelligence platform." www.microsoft.com/en-us/research/uploads/prod/2018/04/BringingAItoBI.pdf.

Ertel, W. 2011. "Introduction." In: *Introduction to Artificial Intelligence.* London: Springer.

Financial statements for Alphabet, Inc. 2021. www.google.com/finance/quote/GOOG: NASDAQ (accessed March 1, 2021).

Financial statements for Amazon, Inc. 2021. www.google.com/finance/quote/AMZN:NASDAQ (accessed March 1, 2021).

Financial statements for Apple, Inc. 2021. www.google.com/finance/quote/APPL:NASDAQ (accessed March 1, 2021).Financial statements for Facebook, Inc. 2021. www.google. com/finance/quote/FB:NASDAQ (accessed March 1, 2021).

Financial statements for Microsoft Corporation 2021. www.google.com/finance/quote/ MSFT:NASDAQ (accessed March 1, 2021).

Financial statements for Netflix, Inc. 2021. www.google.com/finance/quote/NFLX:NASDAQ (accessed March 1, 2021).

Gartner. 2020. "What are Data Science and Machine Learning (ML) Platforms?" www. gartner.com/reviews/market/data-science-machine-learning-platforms (accessed March 1, 2021).

Hernandez, D. 2014. "Meet the Man Google Hired to Make AI a Reality." Wired. www.wired. com/2014/01/geoffrey-hinton-deep-learning/ (accessed March 1, 2021).

Jones, Ben. 2019. "Credit Where Credit is Due: Mary Eleanor Spear." https://medium.com/ nightingale/credit-where-credit-is-due-mary-eleanor-spear-6a7a1951b8e6.

McCulloch, W. and W. Pitts. 1943. "Logical calculus of the ideas immanent in nervous activity." *Bulletin of Mathematical Biophysics,* 5: 115–133.

Netflix. 2009. "Netflix Prize." www.netflixprize.com/index.html (accessed March 1, 2021).

R Core Team. 2020. "R: A language and environment for statistical computing." Vienna, Austria: R Foundation for Statistical Computing. www.r-project.org/.

Raban, D. and A. Gordon. (2020). "The evolution of data science and Big Data research: A bibliometric analysis." *Scientometrics,* 122: 1563–1581.

Sande, G. 2001. "Obituary: John Wilder Tukey." *Physics Today,* 54 (7): 80–81.

The Economist. 2017. "The world's most valuable resource is no longer oil but data." May 6th 2017 edition.

Van Rossum, G. 1995. "Python tutorial." *Technical Report CS-R9526*. Amsterdam, The Netherlands: CWI (Centre for Mathematics and Computer Science).

Veselov, V. 2014. "Computer AI Passes Turing Test in World First." BBC News. www.bbc.co.uk/news/technology-27762088.

Ware, C. 2019. *Information Visualization: Perception for design (interactive technologies)*. 4th edition. Burlington, MA: Morgan Kaufmann.

World Economic Forum. 2018. "Creative Disruption: The impact of emerging technologies on the creative economy." White Paper. www3.weforum.org/docs/39655_CREATIVE-DISRUPTION.pdf (accessed March 1, 2021).

World Economic Forum. 2020. "The Future of Jobs Report 2020." www.weforum.org/reports/the-future-of-jobs-report-2020 (accessed March 1, 2021).

5 Assessing the Potential of Implementing Blockchain in Supply Chains Using Agent-based Simulation and Deep Learning

Mohammad Obeidat and Luis Rabelo

5.1 INTRODUCTION

Complex systems contain numerous components which are distinct but integrated and linked. The entities' relationship within a complex system involves a network of analytical techniques and structures to identify the component information exchange (Weisbuch 2019). Supply chain structures and value chains are highly problematic because of their functions. There are issues with reliability, visibility, and security. Complex structures such as a supply chain are researched using relatively vast amounts of data, which implement various advances and discoveries in how the components and communications function within complex systems.

Blockchain is a new technology that appears to solve security issues in the present supply chain system. Agent-based simulation can model performance improvements due to Blockchain implementation in particular supply chains (Bryk 2017).

5.2 BASIC CONCEPTS

5.2.1 Supply Chain

Supply chains are complex systems because of the number of suppliers, number of customers, number of interactions, conflicting policies, and different decisions and actions. Procurement of raw materials, design, fabrication, technology, network logistics, delivery, and consumer sales make a supply chain a complex system (Law 2017). Supply chain systems and value chains are complex systems due to their nature. Despite proposed research work that attempts to utilize models to solve these

DOI: 10.1201/9781003137993-6

systems' challenges, such as integration and collaboration roles that add value, there are still problems with consistency, transparency, and security.

5.2.2 Blockchain

Blockchain is an evolving technology that has successfully secured financial transactions among trading partners (Ali et al. 2018). A Blockchain system is comprised of a sequence of agreements, a digital network, and protocols between implemented parties (Gupta, Kumar, and Karam 2020). These operations are synchronously registered on the Blockchain, and users can keep track of digital data without the need for centralized monitoring like conventional transactions. A primary function of the Blockchain is to be a centralized database framework (Law 2017). Several research methods have focused mainly on Blockchain's security and privacy issues since its development. When applied to a supply chain, Blockchain technology can enhance transparency and traceability in data and monetary exchanges, building trust and reputation between various entities (Cong and He 2018).

5.2.3 Deep Learning

Deep Learning is a technology that, from Artificial Intelligence and Machine Learning, imposes digital learning computing processes through examples, perhaps with hundreds of different data (McClelland 2017). Deep Learning enables computers to learn information without human intervention. This principle is also called learning by unordered, unmonitored, or unlabeled information. Deep Learning (i.e., a more sophisticated neural network) has interconnected layers that create a relationship between the layers and transfer data from the node of the output values to the receiver node's origin. The more the algorithms learn via large datasets, the stronger they function (Marcus 2018).

5.2.4 Simulation

Simulation is a general term that refers to a mechanism for utilizing a combination of copying and observing a real or predicted physical experiment in a simulated sense. Simulation provides a better comprehension of systems and systems' behaviors and may detect potential issues or inefficiencies (Meng and Liu 2018). Simulation may also distinguish behavior expected, direct or indirect, to enhance or reduce activities or outcomes about a specific process or as a whole system, entity, or occurrence (Anderson 2016).

5.2.4.1 Agent-based Simulation

The agent viewpoint helps one model from the perspective of the users/entities that form the system and understand their individual decisions, actions, and principles of decision making by analyzing the agents' behaviors and actions in the simulation world. Agent-based simulation aids in working with representations of actual, or expected, agent actions rather than abstract versions of behaviors to see a larger version of individual interactions (North and Macal 2007).

5.2.5 Summary of Agents, Deep Learning, and Blockchain

In summary, the topics mentioned above cover the importance of each topic related to this research work. These processes are complicated and need support in decision making. This research comes in place to assess this technology's testing on the supply chain network by using agent-based simulation. Throughout the literature review, we were unable to find any research that assessed the Blockchain's implementation in the supply chains, which led to a research gap.

5.3 PROBLEM STATEMENT AND OBJECTIVE

Supply chains deal with uncertainty, an absence of transparency, lack of information sharing, unpredictability, and the need to create a reputation. Blockchain is crucial because it may solve some of these challenges. However, there is no method to assess if Blockchain will provide advantages to supply chains.

5.4 METHODOLOGY AND FRAMEWORK

The research methodology includes analyzing and constructing a new assessment system to justify Blockchain implementation in supply chain systems. This assessment system uses Deep Learning and agent-based simulation. This research idea originated from a detailed analysis of Blockchain networks and supply chain technologies. This research methodology involves numerous elements, such as research ideas, literature review, gap analysis, system development, case study, process simulation evaluation, thesis validation, and conclusion.

The proposed methodology is mapped to define the respective components, such as the supply chain actors, supply chain functions, Blockchain, and how each of the components interacts with the other. For this purpose, we are using the agent-based simulation environment to model the components mentioned, combined with Deep Learning. Three models are developed to compare the outcomes shown in Figure 5.1: a supply chain without Blockchain and no sophisticated intruder system; a supply chain without Blockchain but using a sophisticated intruder system; and finally, a supply chain using Blockchain and a sophisticated intruder system.

5.5 CASE STUDY

Peer-to-Peer (P2P) platforms like eBay, Uber, and financial institutions like Lending Club (www.lendingclub.com) empower vendors of products and services to negotiate with conventional goods or service providers to sell directly to meet the consumer in a competitive environment without a third-party agent (Aslam and Shah 2017). For example, P2P lending offers customers a virtual credit platform for marketing and promoting their loans at a lower total cost to borrowers or businesses than conventional loan programs. This new business model provides borrowers with added value at lower rates and gives investors the ability to earn attractive or promising returns. Several problems are present in the current processes, such as cybersecurity, transparency, and trust issues.

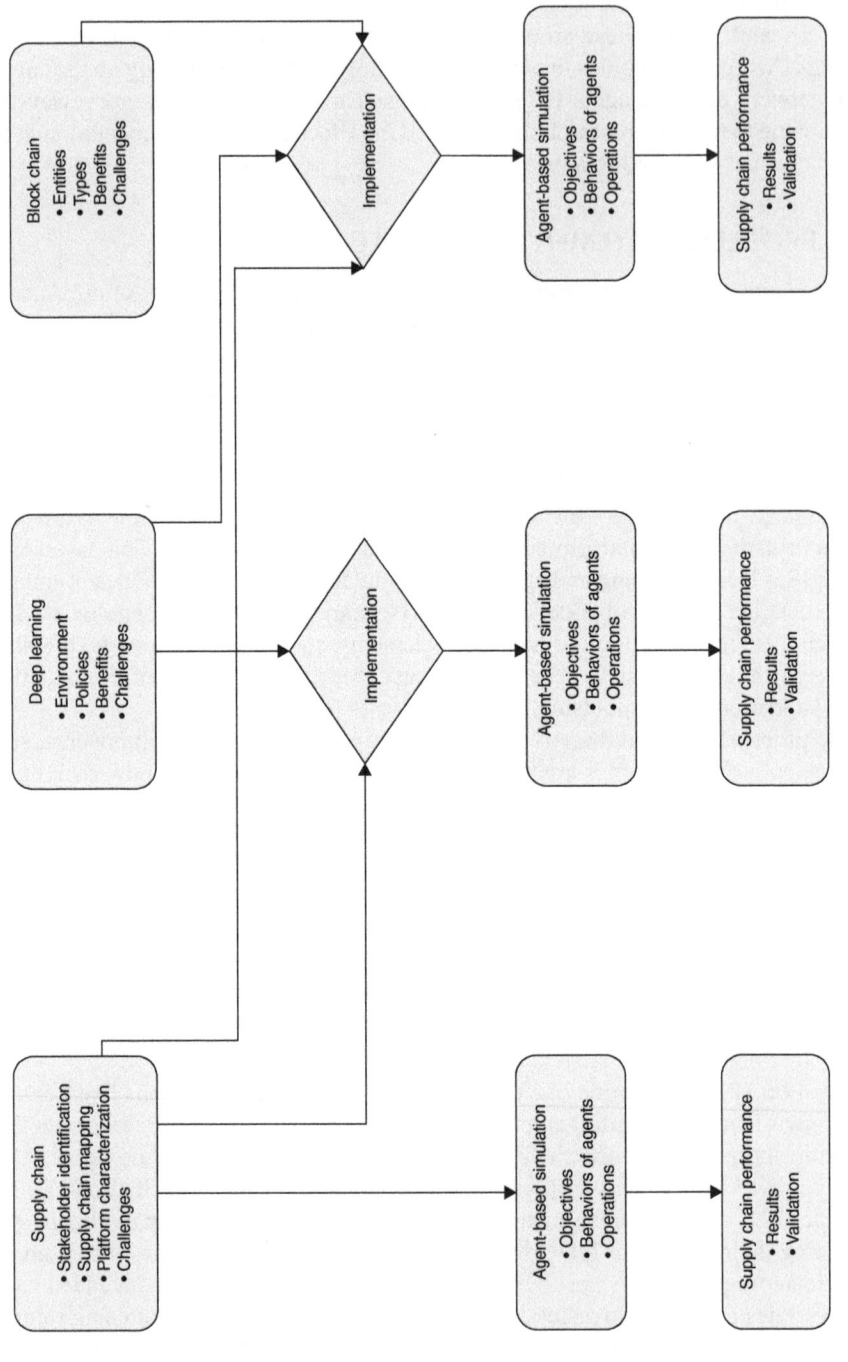

FIGURE 5.1   Three simulation models are utilized to make comparisons

Like other financial firms, P2P lending faces cybersecurity challenges related to financial benefits for attackers who want to intrude and retrieve what they can for a monetary benefit, causing disruptions to these service supply chains (Wang, Greiner, and Aronson 2009). To simulate the P2P lending system's environment, we must familiarize and label the P2P components as agents and simulate them to understand their activities. Figure 5.2 details the process map of the borrower's activities starting from applying for a loan, continuing with the initial screening, and posting the profile to attract lenders. If accepted, the repayment process starts. Figure 5.3 illustrates the process followed by a lender, who starts by setting an account and selecting investments (i.e., providing financial resources to a selected borrower).

Institutions such as the Lending Club face many challenges, such as when the Federal Reserve increases the federal fund rate. Challenges can also be found in trust and transparency, and cybersecurity. This study aims to focus on cybersecurity and transparency/trust issues (Chang 2015).

5.6 IMPLEMENTATION

An *agent-based model* (ABM) is defined by a set of *agent types* with a population each, an environment in which they exist, and finally, interactions between the different agents in the model. It was decided to use the agent-based simulation platform AnyLogic® (www.anylogic.com). In AnyLogic, each agent type is defined by a statechart (i.e., a combination of states and transitions between these states), functions, and attributes. As part of this study, three different agent-based simulation models are developed as follows:

1. Current P2P organization (inspired by the Lending Club).
2. The addition of a Deep Learning IT security system.
3. Same as above with the addition of Blockchain.

5.6.1 CURRENT P2P ORGANIZATION

The initial model has three agents: the borrower, investors, and trust, as shown in Figure 5.4. The interactions among the different populations of agents (i.e., borrowers, lenders, and trust) are also shown in Figure 5.4. Figure 5.5 shows the statecharts for the three agents.

The initial parameter is the agent's class for each of the three statecharts. The statecharts of the borrowers and lenders follow the process maps. However, there are points in the process where cyber attackers can produce changes. The latter may take three values: 1) attacker, 2) suspicious, or 3) normal. Based on organizations such as the Lending Club data, 2% of platform profiles are attackers, 24% are suspicious, and 74% are normal (Lending Club Corporation 2019). Agents of class "suspicious" are not attackers; their profiles raise question marks but end up generally being accepted. To run the simulation model, we must get the distributions for the characteristics based on the actual data from the Lending Club, which provides such information publicly. Those distributions are shown in Table 5.1.

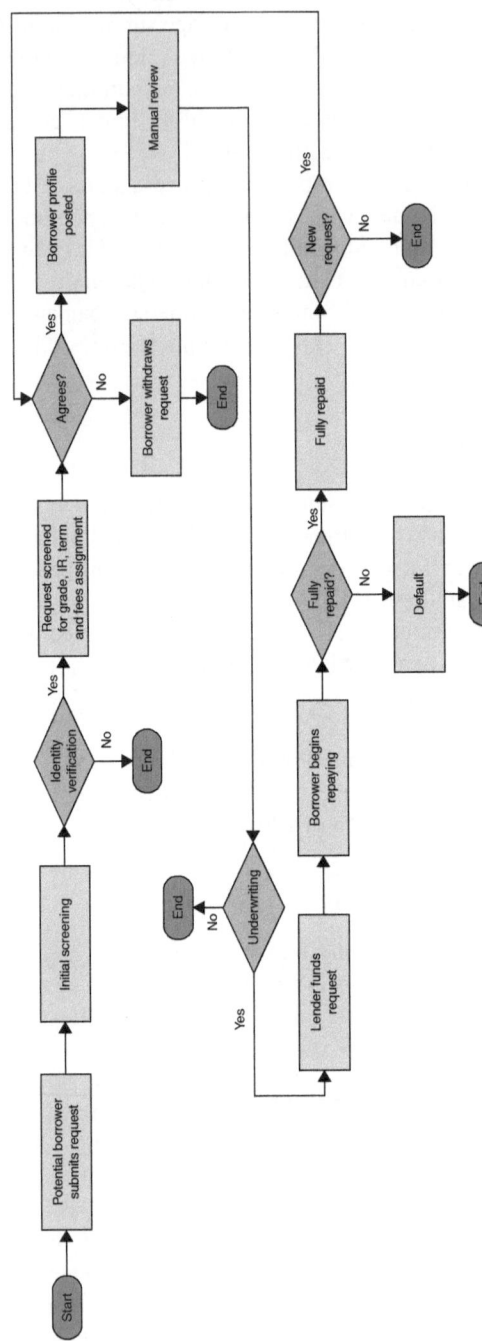

FIGURE 5.2 Borrower activities from the request submission to the repayment of the loan

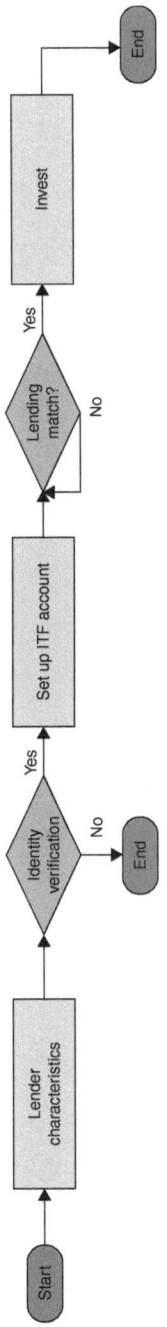

FIGURE 5.3   Lender activities from setting up an account to providing funds to a selected borrower

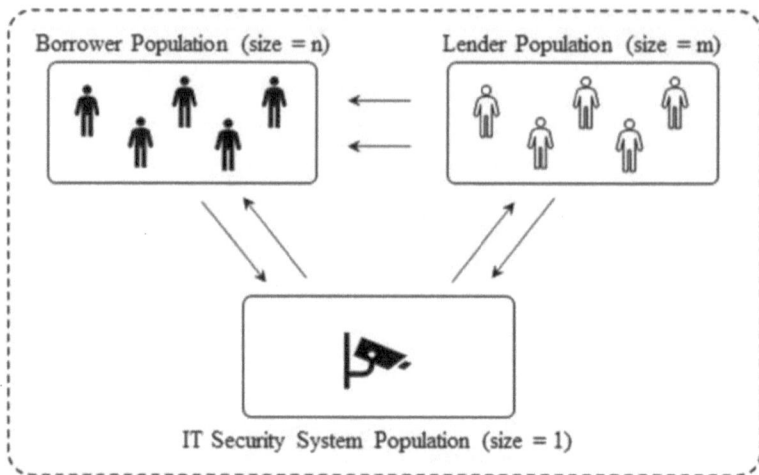

FIGURE 5.4   Current process model

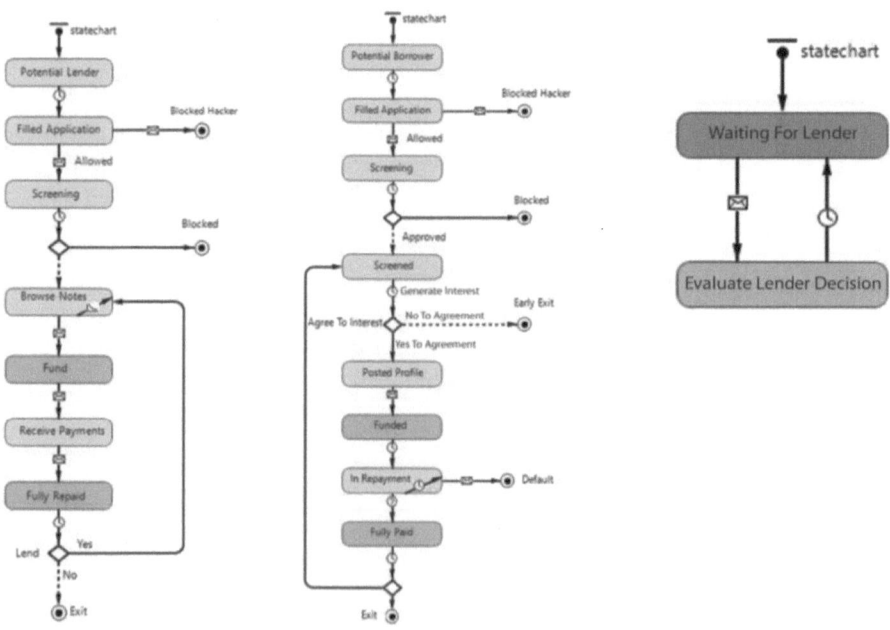

FIGURE 5.5   Lender's, borrower's, and trust's statecharts

TABLE 5.1
Distributions for the Characteristics of the Model

• Loan Amount Custom Distribution	• Term Distribution by Grade
• Borrower Grade Custom Distribution	• Interest Rate Distribution for Grade A
• Interest Rate Distribution for Grade B	• Interest Rate Distribution for Grade C
• Interest Rate Distribution for Grade D	• Interest Rate Distribution for Grade E
• Interest Rate Distribution for Grade F	• Interest Rate Distribution for Grade G
• Originating Fee by Grade	

Source: Lending Club Corporation, 2019.

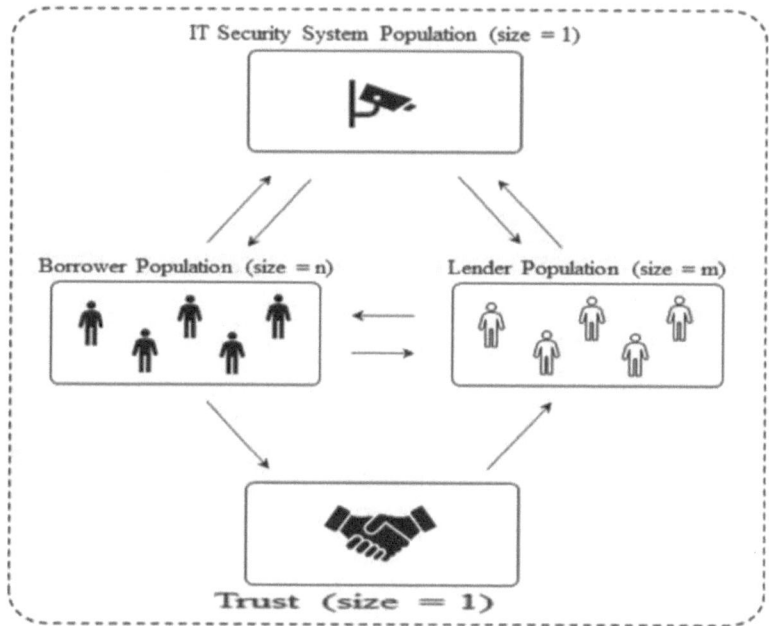

FIGURE 5.6   Upgraded ABM environment and agent interactions

5.6.2 ADDITION OF IT SECURITY SYSTEM MODELED BY USING DEEP LEARNING

The main difference between the previously described model and the one in this section is the addition of a new agent type called the IT security system, which has a population size of one. The interactions between agents, with the addition of this one, are presented in Figure 5.6.

In addition to the statecharts in the first model (see Figure 5.5), the IT security system's statechart is illustrated in Figure 5.7.

The IT security agent is a Deep Learning neural network that is capsulated in the agent-based simulation. The Deep Learning neural network is used to map the

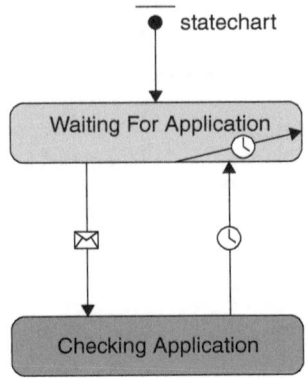

FIGURE 5.7   IT security system statechart

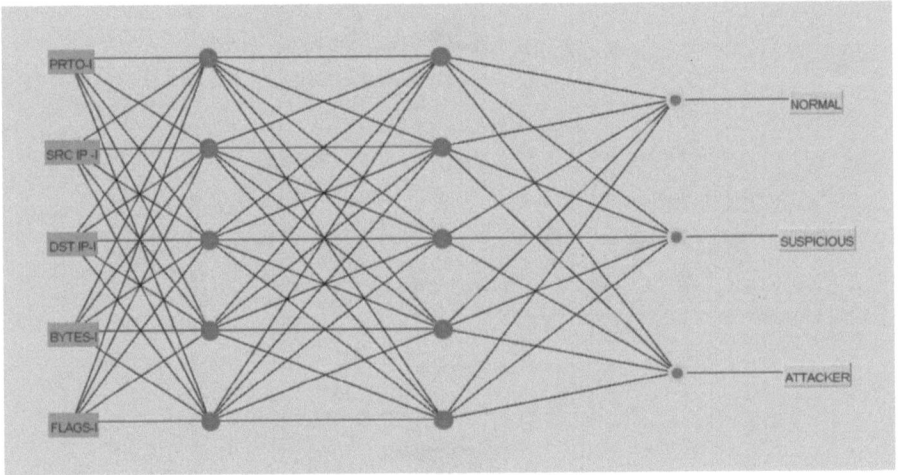

FIGURE 5.8   Deep Learning neural network

characteristics (such as proto, source IP address, destination IP address, number of bytes of the messages, flags raised) of users to the type of users "normal," "suspicious," or "attackers." We used modifications to different datasets and simulations which has similar IT infrastructures to a P2P lending organization. These datasets are CIDDS-001 (Ring et al. 2017). The datasets and corresponding simulations of similar IT infrastructure can generate enough data in the millions of transactions.

KNIME data analytics platform (www.knime.com) is used to build the Deep Learning neural network using Deeplearning4J Integration. Figure 5.8 shows the Deep Learning neural network, which includes the input layer, two hidden dense layers, and the output layer.

Figure 5.9 shows the KNIME flowchart model. This flowchart indicates the data's reading, the process to prepare the data, the partitions of the data, and the

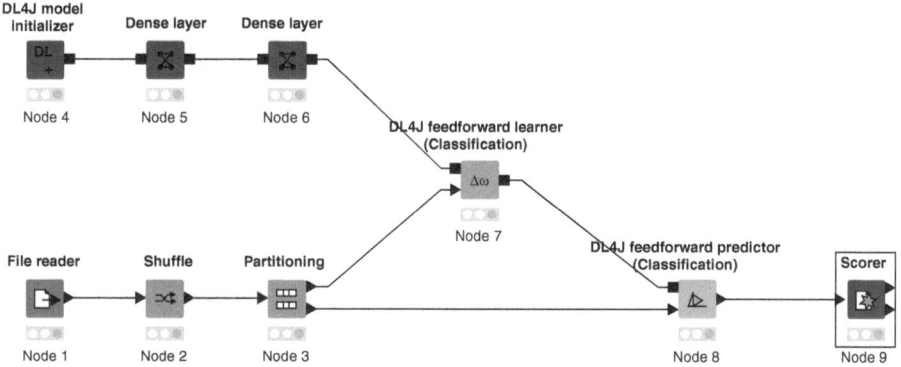

FIGURE 5.9   KNIME flowchart of the Deep Learning neural network

Deep Learning features utilized. The accuracy of this model is 98.539% using testing datasets. The data was divided into three sets: training, validation, and testing. The training dataset was used to get the Deep Learning neural network parameters such as weights. The validation dataset to obtain an appropriate architecture (e.g., number of hidden layers, optimization schemes, etc.). The testing dataset was to test and label the Deep Learning neural network with accuracy.

5.6.3 Addition of Blockchain

The third implementation is to add Blockchain on top of the Deep Learning IT security system. The borrower creates an account and uploads their data, then a block is created for the user and added to the Blockchain, which results in a private key and a public key. The private key is used to access and modify the user data. When the borrower then submits a loan request with his public key, a smart contract is initiated, which holds the request and the borrower information. At this point, the interest rate is calculated, and the lender creates an account. The lender can then view the loan requests and accept or reject them after reviewing the borrower's public key. A smart contract gets executed and withdraws money from the lender account and deposits the money into the borrower account monthly.

5.7   RESULTS

The three scenarios are developed and run, and their results are compared. A fixed number of initial borrowers and investors is assumed to compare results between the different models. One thousand are added to the model, and the model is left running until all agents reach end states. To achieve meaningful results, since the model contains several stochastic components, AnyLogic's Parameter Variation's freeform experiment type is used. This allows to run each of the three models 100 times with different seeds, and the average of the 100 runs is used for analysis. The different scenarios can be interactively chosen through a developed friendly graphical user interface. Moreover, AnyLogic's statistics elements are used to collect the

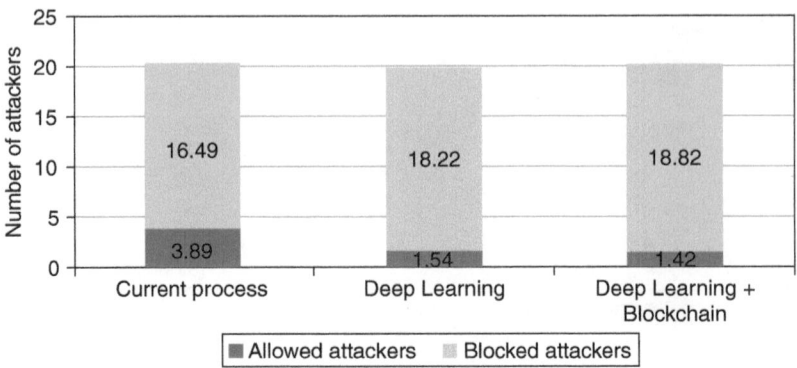

FIGURE 5.10   Number of attackers for each of the simulated environments

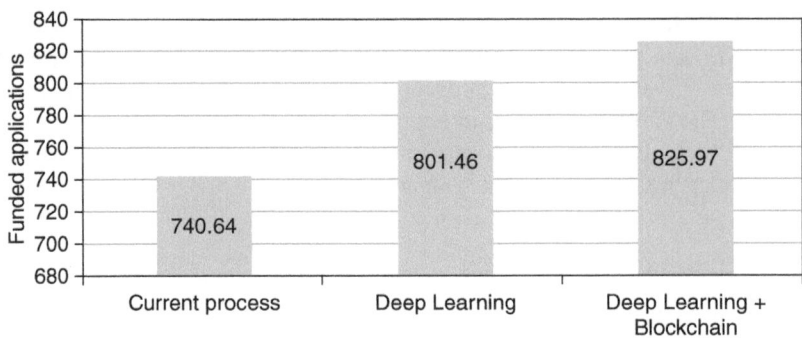

FIGURE 5.11   Number of funded applications for each of the environments

results with statistical information (e.g., mean, standard deviation, mean confidence, etc.). We compared the three models in the following measures: funded applications, blocked/allowed attackers, blocked non-attackers, repeat lenders, and application process duration. The comparisons are shown in Figures 5.10–5.15.

Figure 5.10 shows the number of attackers for each model. The model of the implementation of Blockchain and the IT security system is the one with better performance. The process is enhanced by the added security and the facilitation of the transactions by using smart contracts.

Figure 5.11 displays the number of funded applications. The increase is noticeable for the environment where the IT security system and Blockchain are implemented. The implementation of smart contracts impacts trust, which causes an increment in funded applications.

Figure 5.12 visually explains the reduction of blocked non-attackers. Blockchain creates more trust, and the certifications of actual borrowers help dramatically to identify attackers. Blockchain creates a more efficient process and fewer disruptions to the lending process and a borrower's lifecycle.

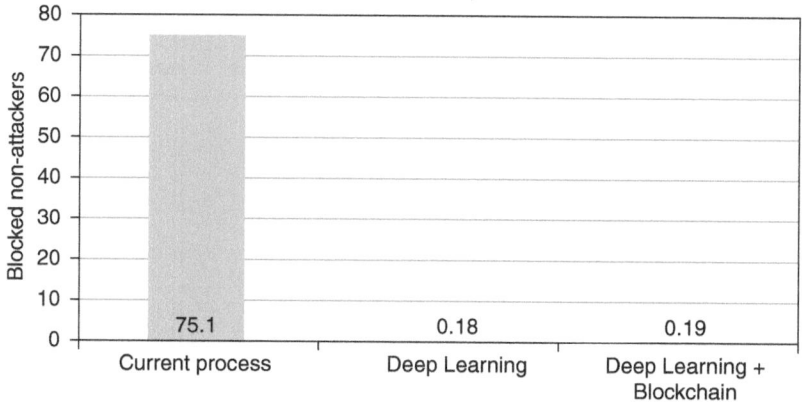

FIGURE 5.12   Blocked non-attackers for each of the environments

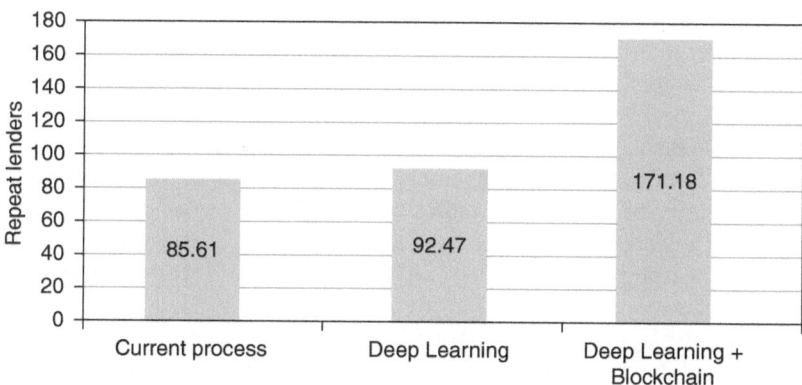

FIGURE 5.13   Repeated lenders for each of the environments

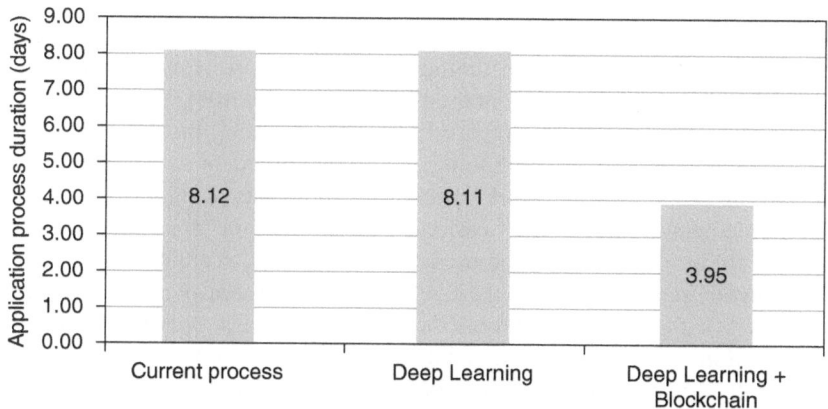

FIGURE 5.14   Duration of the process application in days for each of the environments

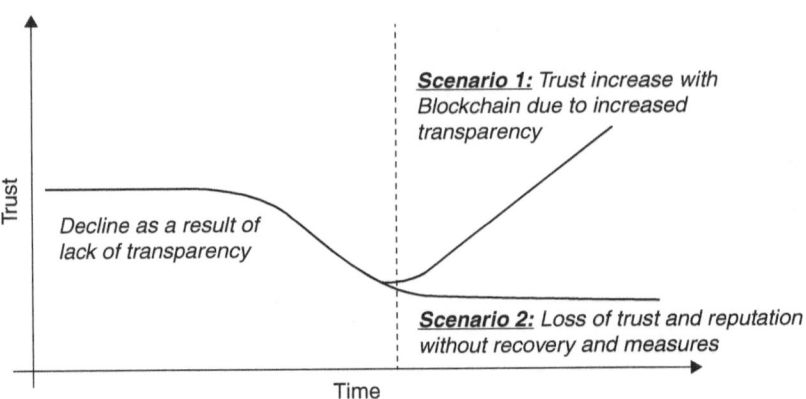

FIGURE 5.15   Trust change over time under different scenarios

Figure 5.13 shows the repeat lenders. The addition of Blockchain and the respective increase in trust help support a virtuous cycle of more investments (i.e., more satisfied lenders who keep investing in the process). This increase is statistically significant.

Figure 5.14 displays the application process durations. The application of Blockchain is one of process improvement. The results are statistically significant, and the superiority of Blockchain indicates spectacular results in the duration of the process by a decrease from 8.11 days to 3.95 days. It is essential to notice that implementing a strong/effective IT security system does not reduce the time of the application process.

Figure 5.15 displays the change in trust for the different scenarios. Smart contracts increase the confidence in the system with the transparency and traceability of the transactions. The different points of the transactions can be verified.

5.8   CONCLUSIONS

Success of the business model is dependent on tangible and intangible aspects. Examples of tangible aspects are number of lenders, borrowers, employees, and services. Whereas examples of intangible aspects are trust, transparency, and reputation. The framework offers a structured approach for integrated design that incorporates process and technologies and stakeholders to complex systems problems. The proposed framework of this study suggests using ABM as a tool to test the implementation of Deep Learning and Blockchain into supply chain and uses a case study. Three simulation models were created to conduct a comparison regarding cyber security and the trust relationship between the involved parties. This research, with its combined tools and advanced topics, proved that the major benefit of implementing Blockchain into supply chains is trust, as it increased between supply chain entities. That conclusion was derived from applying the proposed framework to a case study.

REFERENCES

Ali, M., M. Vecchio, M. Pincheira, et al. 2018. "Applications of Blockchains in the Internet of Things: A comprehensive survey." *IEEE Communications Surveys & Tutorials,* 21 (2).

Anderson, K. 2016. "An Evaluation of Complex Adaptive Evolvable System Simulation." University of Washington. https://digital.lib.washington.edu/researchworks/handle/1773/35501.

Aslam, A. and M. Shah. 2017. "Taxation and the Peer-to-peer Economy." International Monetary Fund. www.imf.org/en/Publications/WP/Issues/2017/08/08/Taxation-and-the-Peer-to-Peer-Economy-45157 (accessed March 1, 2021).

Bryk, A. 2017. "Blockchain: Cyber Security Pros and Cons." www.apriorit.com/dev-blog/462-blockchain-cybersecurity-pros-cons (accessed March 1, 2021)

Chang, W. 2015. "Growing pains: The role of regulation in the collaborative economy." *Intersect: The Stanford Journal of Science, Technology, and Society,* 9 (1).

Cong, L. and Z. He. 2018. "Blockchain disruption and smart contracts." *The Review of Financial Studies,* 32 (5): 1754–1797.

Gupta, S., V. Kumar, and E. Karam. 2020. "New-age technologies-driven social innovation: What, how, where, and why?" *Industrial Marketing Management,* 89: 499–516.

Law, A. 2017. "Smart Contracts and their Application in Supply Chain Management." Massachusetts Institute of Technology. https://dspace.mit.edu/handle/1721.1/114082.

Lending Club Corporation. 2019. "Form 10K." United States Securities and Exchange Commission. www.annualreports.com/HostedData/AnnualReports/PDF/NYSE_LC_2019.pdf (accessed June 8, 2020).Marcus, G. 2018. "Deep Learning: A critical appraisal." New York University. https://arxiv.org/ftp/arxiv/papers/1801/1801.00631.pdf.

McClelland, C. 2017. "The Difference Between Artificial Intelligence, Machine Learning, and Deep Learning." https://medium.com/iotforall/thedifference-between-artificial-intelligence-machine-learning-and-Deep-learning-3aa67bff5991 (accessed April 17, 2018).

Meng, X. and D. Liu. 2018. "GeTrust: A guarantee-based trust model in chord-based P2P networks." *IEEE Transactions on Dependable and Secure Computing,* 15 (1): 54–68.

North, M. and C. Macal. 2007. *Managing Business Complexity: Discovering strategic solutions with agent-based modeling and simulation.* New York: Oxford University Press.

Ring, M., S. Wunderlich, D. Grüdl, et al. 2017. "Flow-based Benchmark Data Sets for Intrusion Detection." www.bibsonomy.org/bibtex/2c92848e1e32fa0a420c477b05a22b4e3/markus0412 (accessed March 1, 2021).

Wang, H., M. Greiner, and J. Aronson. 2009. "People-to-people lending: The emerging e-commerce transformation of a financial market." In *Value Creation in E-Business* Management, edited by M. L. Nelson, M. J. Shaw, and T. J. Strader. Berlin: Springer: 182–195.

Weisbuch, G. (2019). *Complex Systems Dynamics.* Boca Raton: CRC Press.

6 Market Behavior Analysis and Product Demand Prediction Using Hybrid Simulation Modeling

Adalberto Prada and Daniel Ortiz

6.1 UNDERSTANDING THE MARKET AND ESTIMATING PRODUCT DEMAND

For companies, strategic decision making is, in essence, an optimization problem. This optimization problem aims to maximize profit, maximize sales, or a combination of both. As more and better information about market behavior becomes available, the solution to this optimization problem will be closer to the ideal condition, obtaining better benefits and a more significant competitive advantage.

However, given that market conditions change over time and in each place, it is difficult for companies to make decisions about pricing, product specification, or its market positioning strategy. When there is not enough information on market behavior, planning processes are deficient. It generates cost overruns that are usually transferred to price, affecting sales and, once sales have decreased, prices must be reduced, affecting profit. Conversely, a better understanding of market dynamics allows making better decisions, reaching adequate sales and profit result simultaneously.

For example, during 2015, the Colombian currency's substantial devaluation against the U.S. dollar caused a decrease in consumers' purchasing power, reducing domestic consumption and car demand. Devaluation caused prices to increase, and consumers decided to buy cheap cars; that is, in the face of an unfavorable purchasing power scenario, the automotive market became more price elastic, changing preferences regarding car attributes (BBVA Research 2016).

Usually, studies on market dynamics contemplate analysis of demand behavior, supply or price, and the influence of factors that determine economic context. However, factors related to the consumption preferences of individuals are also decisive. In reality, product demand is the sum of purchasing decisions that each person makes, based on their particular situation and interactions with other individuals and the environment. In other words, not only price or market variables affect demand, but consumer behavior also plays a relevant role. It depends on each individual's preference regarding product attributes, socio-demographic conditions (like gender, age, economic status, etc.), the information they receive from other individuals and the

DOI: 10.1201/9781003137993-7

environment, and previous experience with similar products. The purchasing choice of a product will not be given exclusively due to an objective process of evaluating alternatives, but a combination of multiple factors associated with consumers and their environment.

In summary, estimation of product demand cannot be solved solely in a deterministic way. Demand behavior is naturally emergent, random, and multi-variate, and innumerable factors converge both in the context of the market and at the level of consumers and the environment, so it is more difficult for companies to estimate demand under different scenarios. The situation is even more complicated with new products because no historical data allows objectivity in inferring consumers' response.

6.2 MARKETS, COMPLEX SYSTEMS, MODELING, AND SIMULATION

In the economic sciences field, it is common to study variables that affect markets and conclude relationships between them. On the other hand, there are numerous studies to understand consumer preferences. However, it is not usual that, with the systems thinking approach, relationships between market variables are evaluated comprehensively and over time, together with factors that determine consumer behavior.

Difficulties in comprehensively understanding market dynamics are that markets respond to complex systems because their behavior results from multiple interactions and external factors, modifying each element's state of the elements that make up the market. Also, phenomena within markets occur at different levels, which may respond differently to changes in an economic, political, or socio-cultural context.

It is necessary to develop models or representations of elements that compose complex systems and their relationships. Modeling is a way of solving real-world problems when prototyping or experiments are too expensive or even impossible to perform. Since it is not always feasible to develop pure analytical models, simulation techniques such as a discrete event, system dynamics, and agent based are useful. The main difference between these techniques is the approach used to understand the system and capture its structure.

Discrete-event simulation assumes system behavior can be described in states originated from events that occur at specific moments in time. The main components are entities, which are passive elements without interaction, representing people, products, documents, tasks, and messages. These entities can travel through blocks in flowcharts, where they are usually placed in queues to be delayed, processed, harnessed, divided, mixed, or combined (Borshchev and Fillipov 2004).

System dynamics describes system behavior over time, from identifying observable variables and cause–effect relationships. From the concept of feedback widely used in control theory and establishing an analogy with hydrodynamic phenomena, any system can be represented by relationships between variables expressed in terms of stocks, flows that fill or drain these stocks, and parameters that determine the values of the flows (Borshchev and Fillipov 2004). Stocks can represent inventories of materials, populations, accumulated capital, and data storage, among others, and flows can represent information, money, material, people, orders, and capital goods

(Sterman 2000). In mathematical terms, system dynamics models are essentially differential equations systems.

Unlike discrete-event simulation or system dynamics, where variables or elements are studied in an aggregate way, an agent-based simulation system is defined from innumerable interactions and behaviors between the most critical and essential elements of a system: the agents. These agents are explicitly defined, their attributes are assigned to each of them, and behavior patterns concerning other agents and the environment are ascertained.

Simulation paradigms can also be combined to generate hybrid models, which are applied when, due to system complexity, responses are required both at a strategic and operational level (Powell et al. 2015). If the solution does not require a detailed breakdown, the system can be represented in an aggregated form from relationships between variables in a conventional system dynamics model. However, when particularities are also important, it is necessary to associate system behavior at both strategic and operational levels. Using hybrid simulation, phenomena characterized by discrete and continuous responses over time can be studied simultaneously (Mosterman 1999). It is feasible to develop hybrid models that combine system dynamics with discrete-event simulation and/or agent-based simulation.

Hybrid modeling and simulation methodology follows premises used for each specific technique. However, the way the model is built is fundamental. In the bottom–up approach, modeling begins at a very detailed level, applying discrete-event or agent-based paradigms to represent microscopic environments and then applying to consolidate rules to describe general behaviors. On the other hand, the top–down approach begins by modeling the general environment, applying system dynamics, and then descending to the lower layers to represent microscopic phenomena. The way simulations run in a hybrid model is also relevant. Shanthikumar and Sargent (1983) define four types of hybrid models: class I when simulations run alternately and independently; class II when simulations are interconnected and run simultaneously; class III when one of the simulations runs as a subroutine of the other; and class IV when one of the simulations is used to determine the value of a variable or the state of some system element.

Thus, hybrid simulation models can become a vital tool for studying markets. They allow connections of phenomena that occur in an economic context, with those related to consumer behavior and their interaction among themselves and the environment. For example, at the macro-level, system dynamics can describe relationships between variables that define the market and economic environment. In contrast, discrete-event or agent-based simulations can describe consumers' behavior at the micro-level. Figure 6.1 describes market and consumer behavior modeling by the hybrid simulation approach.

6.3 USING SYSTEM DYNAMICS AND AGENT-BASED SIMULATION TO ESTIMATE CAR DEMAND

As an example of applying hybrid simulation models to understand market dynamics and consumer behavior, this section summarizes the most relevant aspects of research performed in 2018. This research project aimed to develop a model applying system

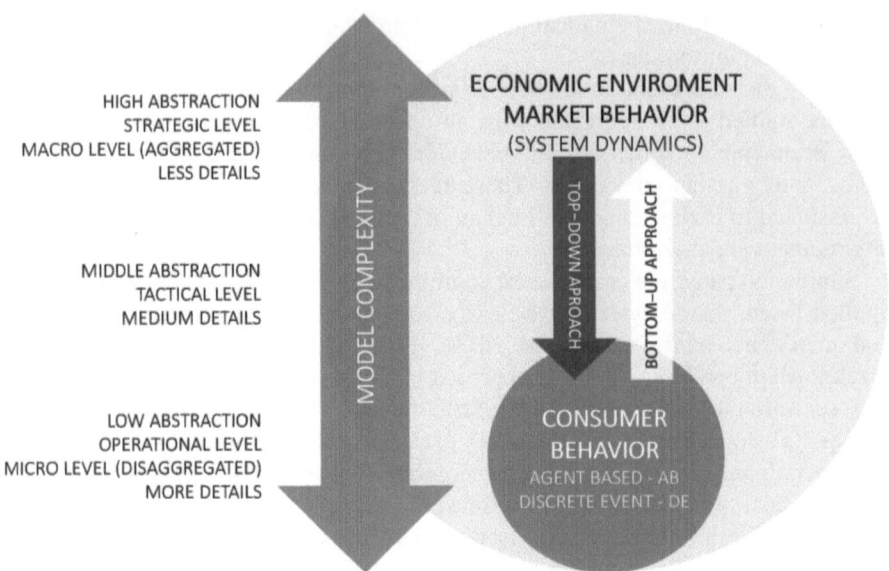

FIGURE 6.1 Market and consumer behavior modeling by a multi-paradigm simulation approach

dynamics and agent-based simulation to predict demand for passenger cars, SUVs, and trucks in Bogotá (Colombia).

This research proposed a new approach to studying market behavior and their consumers, which combines economic and statistical analysis, consumer behavioral models, and complex system simulation paradigms. Unlike traditional economic models, which explain demand based on price and supply, it was suggested that demand is an emergent phenomenon, which depends on factors associated with consumers and their environment. It is bounded by market behavior at the aggregate level. This consumer behavior and demand estimation model represents a paradigm shift and is a starting point for future research in economics and marketing. Another highlight of the research was how the diffusion of cars in the consumer population was proposed. In this case, product adoption was not only based on a combination of adoption or imitation rates. It was also associated with socio-demographic characteristics of individuals and parameters that define the economic environment.

In the same simulation environment, paradigms of system dynamics were applied at the aggregate level to describe the behavior of variables that influence supply, price, and demand; and simultaneously, an agent-based simulation was used at the disaggregated level to model consumer behavior. Applying this modeling strategy was possible to estimate car demand and obtain information on car buyers' characteristics in each period.

Additionally, thanks to the system's thinking approach, all factors that have a relevant impact on car demand behavior were identified and analyzed, including variables associated with markets of substitute and/or complementary products. Car demand was estimated, and motorcycle demand, used-car demand, and other associated variables. In this way, it was found that new-car demand contributes to used-car demand (and vice versa). On the contrary, an increase in motorcycle demand reduces sales of new and used cars.

Although modeling was carried out under two different perspectives, they were interconnected. This characteristic of the model was the key to understanding the economic environment's influence on car consumers' purchase decisions. For example, if economic conditions are not favorable, demand will be affected. Therefore, there will be fewer adopters of an evaluated product; this situation limits its diffusion, reducing or increasing potential adopters' population. Thanks to this approach, it was possible to know that representative market rates and per capita income significantly impact car demand. Still, the market does not react immediately to economic environment changes.

This hybrid simulation model (system dynamics–agent-based) achieved high-accuracy forecasts for new-car demand in Bogotá, with a deviation from real data of less than 1%. Furthermore, the simulation model allowed us to understand the influence that changes in the economic environment have on automobile market behavior, the sensitivity of demand to price and other car attributes, and the effect of word of mouth and advertising on consumer decisions.

6.3.1 Modeling Market at the Aggregate Level (System Dynamics)

The market was modeled as an inventory structure at the aggregate level. New car registrations flow, increasing car fleet (like an inflow to a stock), while scrapping drains it. Car price and the rest of the variables act as regulating factors of these flows and stocks. Figure 6.2 shows the modeling strategy at the aggregate level.

The automotive market comprises a widely diversified product portfolio. In other words, unsatisfied demand situations are rare and isolated and do not affect the market's global dynamics. Registrations and demand have been considered as equivalent variables. In this way, the flow of registrations or car demand also determines the total number of buyers who adopt a new car in a certain period. In

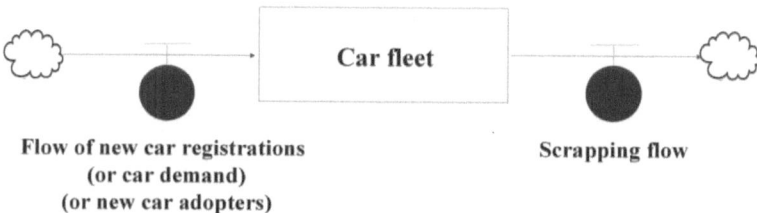

FIGURE 6.2 Dynamic strategy for system modeling at the aggregate level, where car fleet is a stock

contrast, a car fleet defines the total number of cars to be distributed among the population of potential buyers.

Beyond micro-economic considerations, under a systemic approach, system
behavior will depend on the interaction of variables beyond demand, supply, and
price. These variables are considered exogenous and are mainly associated with
the macro-economic environment. On the contrary, price, demand, and supply are
considered endogenous since they can be determined from system behavior and the
influence of exogenous variables. The system will be represented in conjunction with
parameters that define the macro-economic context and market variables.

Thirty-five indicators and variables were defined initially for the system dynamics
model. These indicators and variables are used in different publications and studies
of the automotive industry. After analysis, 18 were finally selected. Figure 6.3 shows
the selected variables from the exogenous and endogenous viewpoints. Variables
associated with motorcycles and the used-car market were included because they
have a relevant impact on car demand behavior. From the information available in
secondary sources, historical data for each of these variables between 2007 and 2016
was obtained. These data served as the basis for statistical analysis and mathematical
models.

Ishikawa diagrams were developed to formulate possible interactions and
cause–effect relationships between system variables. Subsequently, to avoid any
bias in identifying logical cause–effect relationships, correlation and causality tests
were performed, according to criteria established by Pearson (1895) and Granger
(1969). Figure 6.4 shows the Ishikawa diagram built, which confirmed and discarded
causalities for all variables regarding new-car demand. This analysis was repeated for
each one of the variables.

Causal diagrams were constructed from the confirmed causal relationships for
each variable, with signs of causalities ("polarities") and feedback loops that arise
from relationships between endogenous and exogenous variables, as displayed in
Figures 6.5 and 6.6.

Forrester diagrams were constructed for the system at the aggregate level from
confirmed causalities and delays, as shown in Figure 6.7, which details two stocks.

Exogenous variables (Economic context)		Endogenous variables (Automotive market)	
	Related to demand	Related to supply	Related to Price
1. Per capita income	7. New-car demand	15. Car supply	17. Average car price
2. Bank interest	8. Used-car demand	16. Motorcycle supply	18. Average motorcycle price
3. Unemployment	9. Car fleet		
4. Representative market rate	10. Car scrapping		
5. Inflation	11. New-motorcycle demand		
6. Consumer confidence	12. Used-motorcycle demand		
	13. Motorcycle fleet		
	14. Motorcycle scrapping		

FIGURE 6.3   Selected variables for the model at the aggregate level

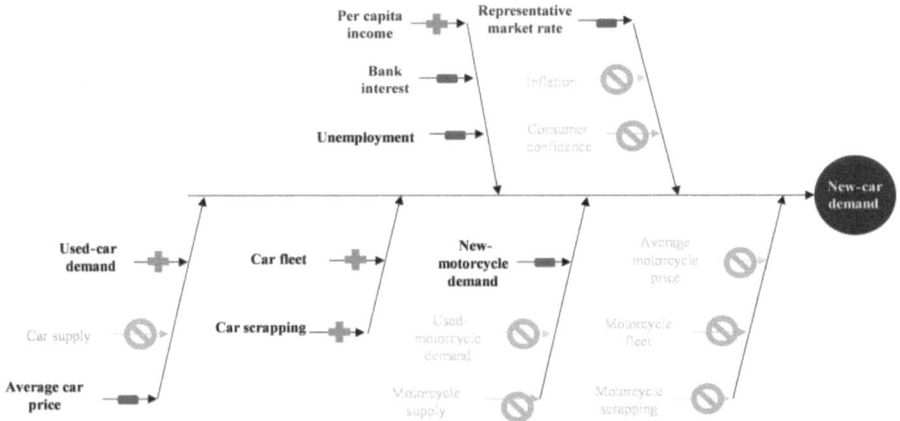

FIGURE 6.4 The Ishikawa diagram discarded and confirmed causal relationships between variables at the aggregate level

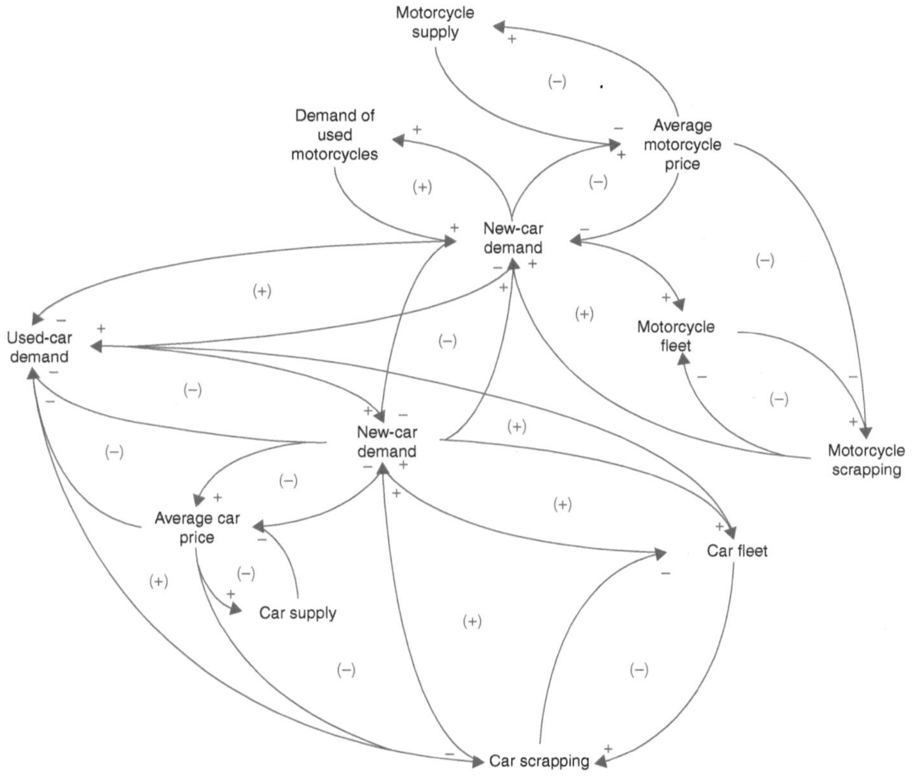

FIGURE 6.5 Causal loop displaying the causal relationships between exogenous and endogenous variables

Note: New-car demand has fascinating relationships with "Average motorcycle price" and "Average car price."

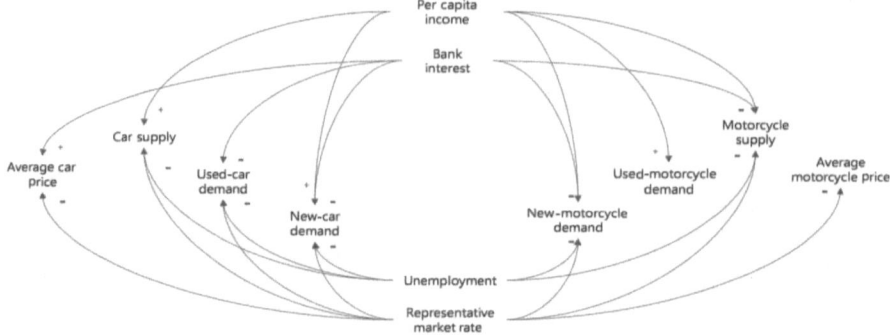

FIGURE 6.6 Causal diagram of the system at the aggregate level

Note: This causal loop shows the importance of "Per capita income" in the decision-making process.

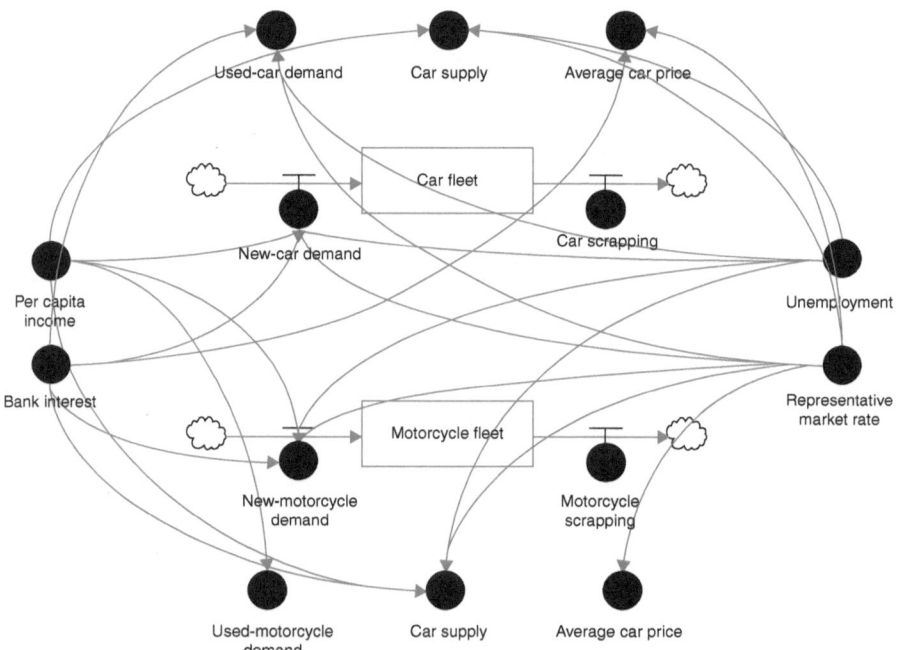

FIGURE 6.7 "Car fleet" and "Motorcycle fleet" are two stocks of the model

Therefore, the system of equations that describes market dynamics at the aggregate level was defined as:

$$P_{AV}\left(t\right) = P_{AV}\left(t_0\right) + \int_{t_0}^{t} \left[M_{VN}\left(t\right) - D_V\left(t\right)\right]dt \qquad (1)$$

and,

$$P_{AM}(t) = P_{AM}(t_0) + \int_{t_0}^{t} \left[M_{MN}(t) - D_M(t) \right] dt \tag{2}$$

Where:

$P_{AV}(t_0)$: Car fleet, at time $t = 0$
$M_{VN}(t)$: New-car demand
$D_V(t)$: Car scrapping
$P_{AM}(t_0)$: Motorcycle fleet, at time $t = 0$
$M_{MN}(t)$: New motorcycle demand
$D_M(t)$: Motorcycle scrapping

It was also validated as established by time-series regressions, finding that they are stationary in mean and variance—this attribute allowed to build a system of equations from a model of auto-regressive vectors (VAR). Applying time-series regressions, new-car demand, and other system variables, can be expressed as:

$$Y(t) = \beta_0 + \sum_{i=1}^{m} \sum_{j=1}^{n} \beta_{ij} \Delta X_{i_{t-j}} \tag{3}$$

Where:

Y : Dependent variable
β_0 : Constant parameter
m : Number of independent variables
n : Optimal number of lags (or delays)
β_{ij} : Coefficient of the independent variable X_i , for lag j
$\Delta X_{i_{t-j}}$: First difference of the independent variable X_i , at time $t - j$

Separating sums corresponding to variables associated with economic context (exogenous variables) from those of the automotive market (endogenous variables), we have:

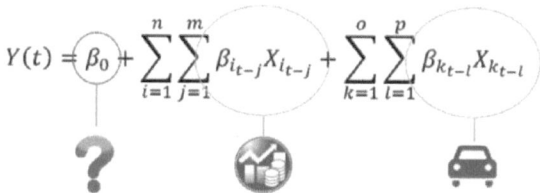

$$Y(t) = \beta_0 + \sum_{i=1}^{n} \sum_{j=1}^{m} \beta_{i_{t-j}} X_{i_{t-j}} + \sum_{k=1}^{o} \sum_{l=1}^{p} \beta_{k_{t-l}} X_{k_{t-l}}$$

VAR analysis allowed consolidation in the constant parameter β_0 those phenomena that were not observed or that have been discarded because they are considered unrepresentative.

6.3.2   Modeling Market at the Disaggregate Level (Agent-based)

At the disaggregated level, the agent-based simulation model describes how consumers make purchase decisions due to the conjunction of external stimulus and intrinsic factors associated with each consumer (or agent in the model). External stimuli refer to information that consumers receive as input for decision making, associated with word of mouth, advertising, and perceptions about the environment. Intrinsic factors are the attributes of each individual, such as age, gender, marital status, parental status, socio-economic stratum, educational level, and occupation. Also added to the intrinsic factors were ownership of a car and/or motorcycle.

The system was representative of a reduced portion of potential consumers; that is, each agent was associated with groups of individuals with the same characteristics and therefore following the same behavior patterns. The sum of behaviors of all agents in the model represents the entire potential consumer population's behavior. This approach reduced the complexity of the model significantly. Figure 6.8 displays the characterization of an agent.

In addition to attributes associated with socio-demographic conditions and ownership of a car or motorcycle, agents were endowed with memory to retain or forget messages that they received about an evaluated car. Messages received by word of mouth always remained current and available in the memory of agents for making purchase decisions. On the other hand, messages received by advertising, unless they were reinforced, would be forgotten as time progressed. Figure 6.9 depicts the variables that describe intrinsic factors and external stimulus.

A survey was conducted among 1,065 inhabitants of Bogotá who met potential car consumers' characteristics to obtain information about car purchase preferences and consumer attributes. This information served as the basis for statistical analysis and construction of a mathematical model at the system's aggregate level. Figure 6.10 shows the car attributes and socio-demographic conditions asked in the car consumer survey in Bogotá.

Behavior rules for agents were established based on the contributions of Schiffman and Kanuk (2014) and Rogers (1983) about consumer behavior and diffusion of innovations models. The adoption process begins with acquiring a new car, which depends on recognizing the need, external stimulus associated with the environment's

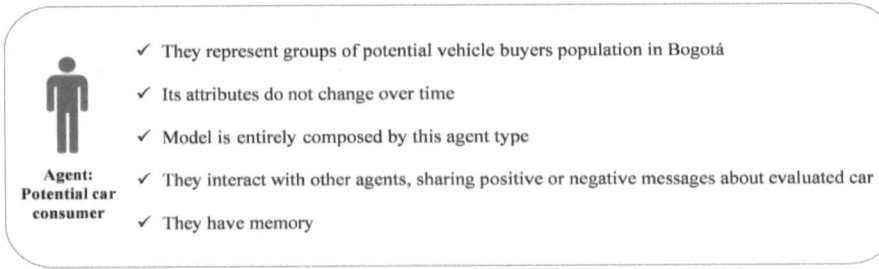

FIGURE 6.8   Characterization of an agent in the model

![Agents] **Agents**	![Environment] **Environment**	
Intrinsic factors	**External stimuli**	**Vehicle attributes**
1. Gender	11. Advertising	14. Price
2. Marital status	12. Positive word of mouth	15. Body type
3. Parental condition	13. Negative word of mouth	16. Colour
4. Age		17. Engine displacement
5. Occupation		18. Fuel type
6. Education level		19. Transmission type
7. Socioeconomic strata		
8. Vehicle ownership		
9. Motorcycle ownership		
10. Memory		

FIGURE 6.9   System variables at the disaggregated level

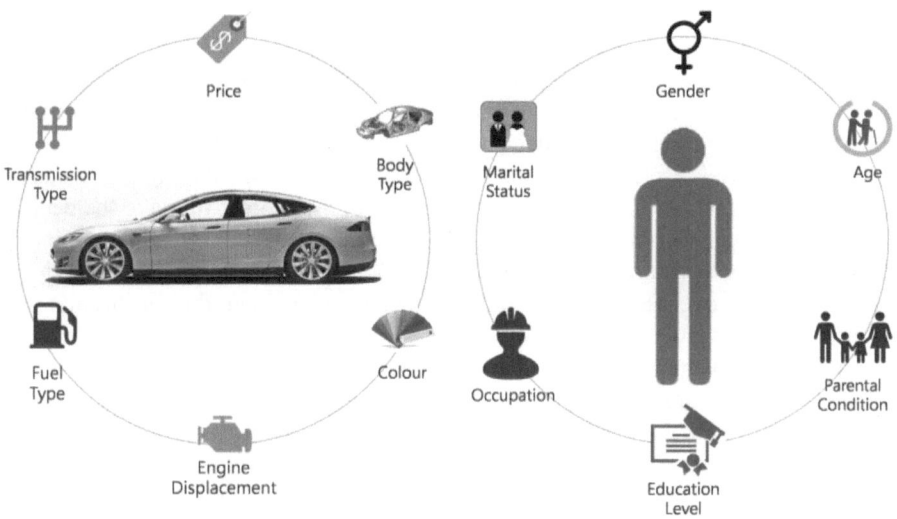

FIGURE 6.10   Car attributes and socio-demographic conditions queried in the car consumer survey

perceptions, and intrinsic factors. Based on this, a consumer will decide whether to buy a car if the environment is conducive and meets conditions.

After the consumer decides to purchase a new car, he/she must decide which product to purchase. All consumer behavior models agree that it is essential for buyers to know the product. That is, what is not known is not acquired! Even if this knowledge has been acquired only a few moments before making a decision. Word of mouth may also be harmful, which implies that communication between consumers

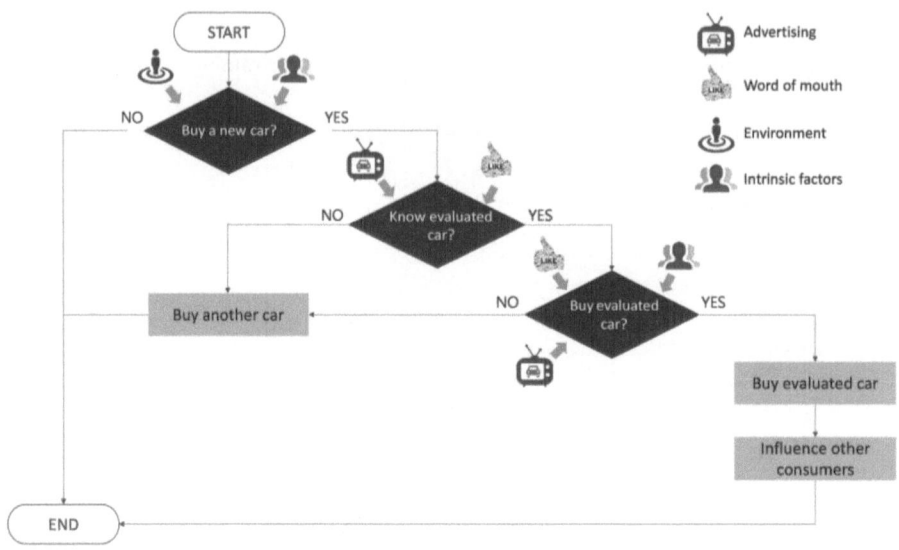

FIGURE 6.11 Decision-making process proposed for the model

may favor or harm a purchase decision. For example, if a product has had technical failures, consumers will be detractors of the product. They may send negative messages to other consumers, causing them to choose a different alternative. If people have not received information about the product through word of mouth or advertising, it is understood that they do not know it and therefore will make the decision to buy another car. This approach recognizes the importance of product dissemination through different promotion channels and recognizing those who have previously acquired knowledge. However, in addition to knowledge about the car through word of mouth and advertising, consumer's choices will also depend on intrinsic factors.

Unlike a conventional diffusion model, in this case product adoption does not depend on fixed rates of innovation and imitation. Production adoption depends on speed and effectiveness: advertising spreads and positive word of mouth of product, presenting a favorable economic environment that maintains or increases the number of potential buyers, and intrinsic factors associated with each of them. For example, if macro-economic conditions change, demand will be affected, and buyers' population (or potential adopters of the product) will decrease. If the product has a bad reputation and negative word of mouth increases, there will be fewer adopters. If the number of contacts of advertising messages increases, the number of adopters will increase. Figure 6.11 shows a simplified diagram of the described decision-making process.

In summary, consumers or agents in the model perform the following tasks:

- **First task:** Decide whether or not to buy a new car.
- **Second task:** Decide whether or not to buy the car that is being evaluated.
- **Third task:** Share messages associated with the evaluated car to other people.

6.3.3   Integration of Simulation Paradigms

Once variables, parameters, interactions, and behavior rules were defined, a hybrid simulation model was developed by integrating system dynamics with agent-based simulation. The connection between both simulation paradigms occurs through the flow of new-car demand, which is equivalent to the number of consumers who choose to buy a new car in each period. Considering definitions of Swinerd and McNaught (2011) and Shanthikumar and Sargent (1983), the developed model can be characterized as follows:

- Built in a top–down approach, modeling begins at the aggregate level and then descends to the disaggregated level (agent-based).
- Designed as an integrated model, since the simulations run in parallel over time, and there is a flow of information from the aggregate level to the disaggregated level.
- Composed as a model of stored agents, considering that the system dynamics model variables are used to limit the number of agents that could adopt the evaluated car.

Figure 6.12 shows a simplified diagram of the integrated model.

6.3.4   Simulation Runs

Simulation runs were performed using AnyLogic® multi-paradigm language (www.anylogic.com). In an AnyLogic programming environment, simulation contemplates a single type of agent: potential buyers. This agent has been endowed with gender, age, marital status, parental status, educational level, socio-economic stratum, occupation, owner of motorcycle or car, memory, and the condition of being a detractor, neutral, or promoter. Simulation is executed with a reduced group of agents representing the entire consumer population. However, because it is not feasible to know precisely the combination of each potential consumer's attributes, attributes associated with intrinsic factors were randomly assigned.

The simulation begins with a panel where the user inputs initial parameters that are adjustable; also, users can select attributes of the car to be evaluated, frequency and scope of advertising, and the expected percentage of promoter and detractor consumers. As it is an agent-based simulation model, it is possible to identify agents who adopted the car in each run. This information is critical because not only is the number of adopted cars known, but also buyers' attributes. In addition to new-car demand, the system dynamics model allowed to estimate the rest of the variables considered at the aggregate level of the system, including demand for motorcycles and used cars, the average transaction price of new cars, the fleet of cars and motorcycles, and the scrapping rates of cars and motorcycles. As an example, results obtained in a simulation run carried out for the year 2017 are presented in Figure 6.13.

For this example, the attributes used regarding a particular car to be evaluated are depicted in Figure 6.14.

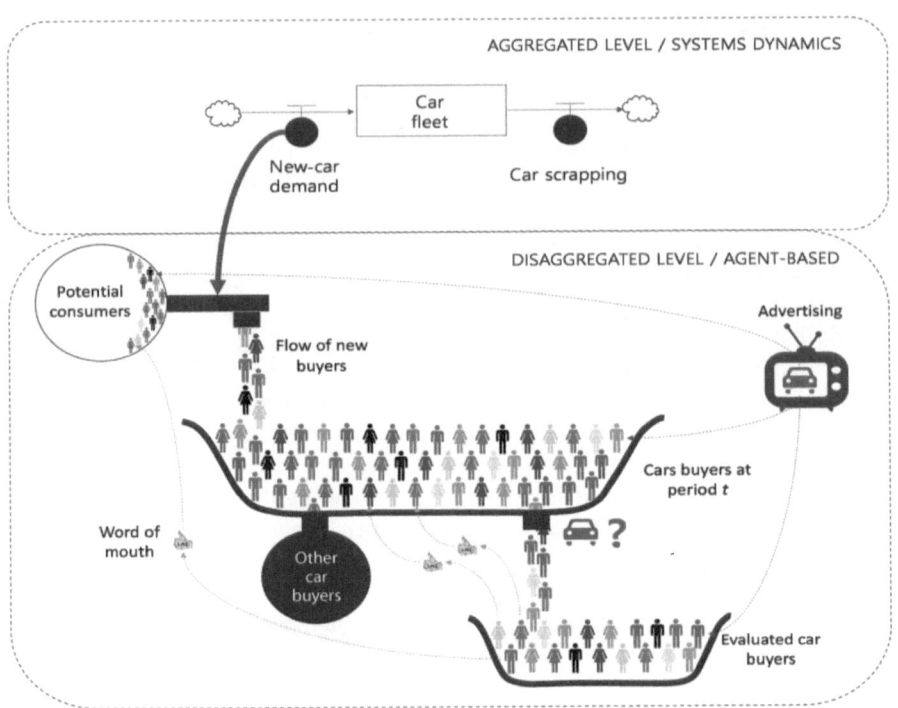

FIGURE 6.12 Simplified illustration of the hybrid simulation model—system dynamics determines the flow of new buyers, while agent-based determines the number of buyers who choose to evaluate a car

Initial Parameter	Criteria
Economic context	Macro-economic conditions presented in 2017
Vehicle advertising strategy	Scope of promotions: **20% of total consumers**
	Promotion frequency: **1 per year**
Determinants of word of mouth	Percentage of promoters: **30%**
	Percentage of detractors: **5%**
Number of agents used in simulation	**10,000 agents**

FIGURE 6.13 Results obtained in a simulation run, 2017

Figure 6.15 shows the user interface for the simulation run, and Figure 6.16 displays the results using AnyLogic.

Applying these premises shows that 92 agents chose to buy a new car during 2017, which is equivalent in the entire system to 71,117 demanded units of new cars

Image	Price	Body type	Colour	Engine displacement	Fuel type	Transmission type
	COP 60 million	Four door sedan car	Gray	Between 1500 cm³ and 2000 cm³	Gasoline	Manual

FIGURE 6.14 Attributes of a particular car for a simulation run

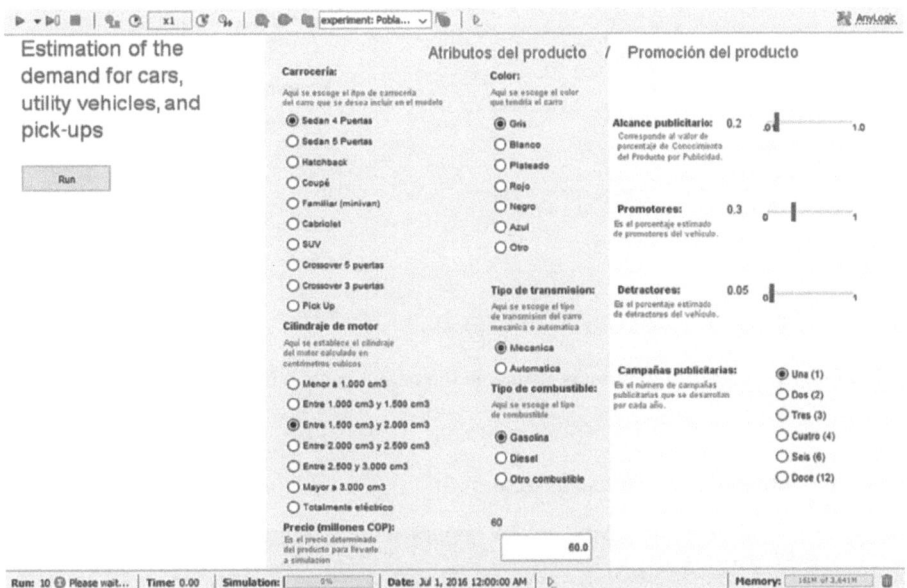

FIGURE 6.15 AnyLogic® user interface; input data panel (initial parameters) for the simulation run

throughout this year. Nine agents also decided to buy the evaluated car, representing in the real system 4,977 demanded units. Additionally, it was known that 55% of adopters of the evaluated car are men, 89% have children, 78% have a university degree, and 67% already had a car before this purchase, among other relevant characteristics. Figure 6.17 displays the attributes of vehicle adopters.

6.3.5 MODEL OPTIMIZATION

6.3.5.1 The Optimal Number of Simulation Runs

Considering probability distributions have been associated with agents' variables and rules, the model response will present a random behavior; therefore, there will be as many results as simulation runs are performed (considering the same input parameters). According to this, each run's expected value will be within a specific confidence interval, defined from the corresponding mean and standard error. For this

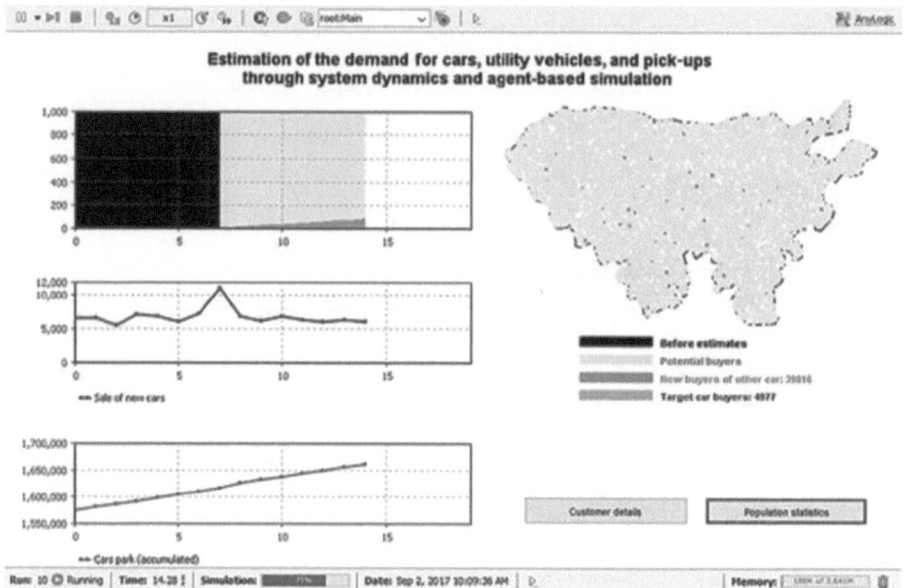

FIGURE 6.16 View of the primary results in the output panel built using AnyLogic®

Attributes of vehicle adopters

Agent code	Gender	Marital status	Parental condition	Age	Education level	Stratum	Occupation	Vehicle ownership	Motorcycle ownership
396	Male	Married	With children	34	Graduate	3	Employee	No	No
1238	Male	Single	Childless	45	Graduate	5	Employee	Yes	No
1362	Female	Single	Childless	25	Graduate	4	Employee	Yes	No
1470	Male	Single	With children	33	Graduate	2	Employee	No	No
2941	Male	Married	With children	43	Graduate	4	Employee	Yes	No
2979	Female	Married	With children	27	Non graduate	5	Employee	Yes	No
5448	Male	Single	With children	48	Non graduate	4	Employee	Yes	No
5750	Female	Married	With children	29	Graduate	4	Employee	Yes	No
9087	Female	Married	With children	34	Graduate	4	Employee	No	No

FIGURE 6.17 Attributes of adopters who have evaluated the car

case, an optimal number of runs was established based on the premise of obtaining a confidence level of 95% and a standard error of 10% respect value of the mean. The authors determined that it is necessary to perform at least 50 simulation runs to obtain the required confidence and standard error level.

6.3.5.2 The Optimal Number of Agents

It could be thought that modeling with a number of agents equal to the total population is ideal. However, this condition is inefficient because it requires a large data processing capacity, increasing when simulation runs are executed. But modeling with a minimal number of agents is also problematic, because it implies that each agent must represent many consumers in the system, consequently decreasing the model's ability to describe small variations. For example, suppose a population of 50,000 agents is considered. In that case, each of them will represent 111 real consumers (considering the total population of consumers in Bogotá). That is, variations in the number of buyers will only be detected in the same magnitude. Conversely, if only 1,000 agents are chosen, the model's detectable changes will be to ± 5,528 buyers. In conclusion, modeling with the wrong number of agents will affect the likelihood outcomes of predictions.

To estimate the optimal number of agents that this model required, the authors obtained calculated error when simulating with a number of agents and found that error has a potential behavior as βX^{α} where $\alpha < 0$, $\beta > 0$, converges to 0 as the number of agents considered in the simulation increases.

Figure 6.18 shows the error in predicting new-car demand during the year 2017 for different numbers of agents used in the simulation runs. In this case, the simulation runs with 10,000 agents can be accepted since the error obtained is less than 5%. However, if it is necessary to estimate a lower demand, the number of agents must be greater to avoid very high deviations. For example, if a population of 20,000 agents is considered, estimates of 19,000 buyers per year can be made, with an error of less

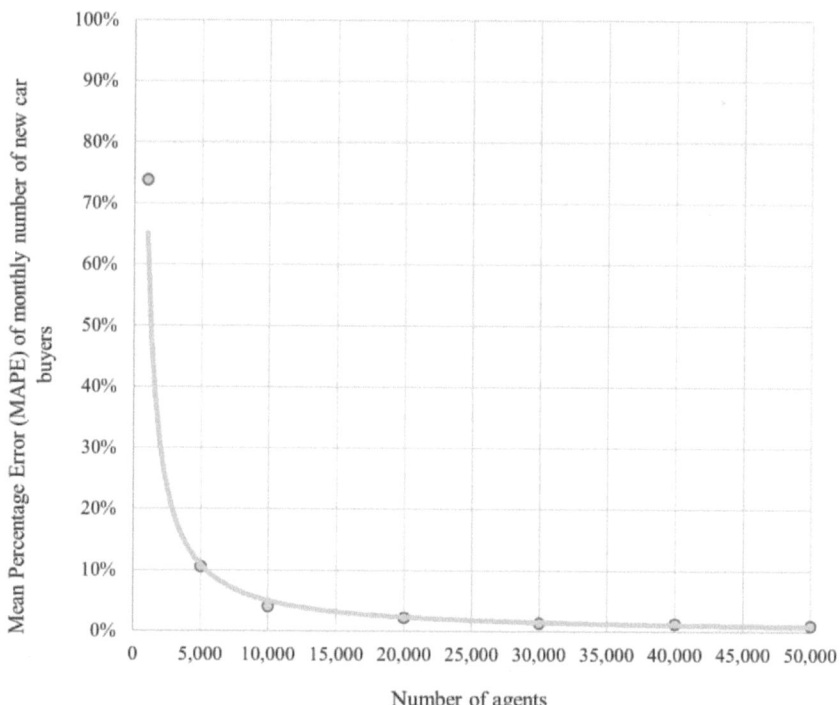

FIGURE 6.18   Prediction error of new-car demand for different numbers of agents during 2017

than 10%; if the population is 50,000 agents, estimates of 8,000 buyers per year can be made with the same error rate. With a population of 350,000 agents, estimates of even 800 buyers per year could be made, keeping errors less than 10%. This result is significant because it shows that it is unnecessary to model the entire population to obtain reliable forecasts if the model is designed correctly.

3.5   MODEL VALIDATION AND SENSITIVITY ANALYSIS

Estimations for all selected variables were compared with real market data to validate the hybrid model. As the model was built from historical data series up to December 2016, validation tests consisted in determining deviations of model predictions as respect to real data obtained in 2017, considering the same economic context (i.e., the same input parameters). The following errors were obtained when running simulation and comparing data, as displayed in Figure 6.19.

In conclusion, the model could predict car demand in Bogotá with less than 1% error. Excellent predictions were also achieved for the rest of the system variables. Additionally, the model allows evaluating car demand sensitivity to changes in the input parameters. Regarding parameters that define economic context, model experiments showed new-car demand reacts positively to an increase in GDP per capita. Conversely, car demand reacts negatively if unemployment or bank interest increase. Figure 6.20 shows how changes in these initial parameters affect estimated demand for the year 2017.

6.4   CONCLUSIONS

Hybrid simulation modeling is an essential methodology in predictive analytics. Hybrid modeling goes beyond the traditional models of regression. System dynamics models are effective at the aggregate level, and agent-based modeling is a robust

Variable		Real value	Model estimation	% error
Car market	**Demand for new cars**	**77,115**	**77,020**	**0,1%**
	Demand for used cars	250,950	254,181	1,3%
	Vehicle supply	87,309	87,450	0,2%
	Average vehicle price	COP 46,481,609	COP 45,750,217	1,6%
	Growth of vehicle fleet	67,891	67,669	0,3%
	Scrapped vehicles	9,224	9,351	1,4%
Motorcycle market	Demand for new motorcycles	32,6%	9,0%	0,6%
	Demand for used motorcycles	11,3%	4,2%	0,7%
	Motorcycle supply	24,3%	8,1%	0,6%
	Average motorcycle price	3,4%	0,8%	0,4%
	Growth of motorcycle fleet	6,576	6,536	0,6%
	Scrapped motorcycles	989	1,071	8,3%

FIGURE 6.19   Model validated at the aggregate level

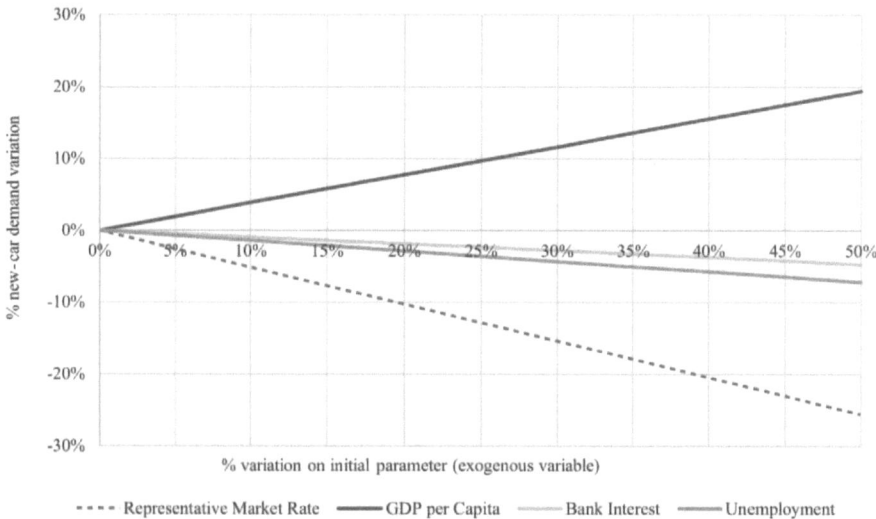

FIGURE 6.20 Sensitivity of new-car demand to changes in the exogenous variables

scheme at the disaggregated level. Some specific conclusions from our study of Bogotá are:

- **The market is sensitive to changes in the economic context:** Vehicle registrations and other endogenous variables are sensitive to changes in the economic environment, such as unemployment and bank rates.
- **High incidence of the exchange rate and per capita income in vehicle demand**: It was found that the factors with the most significant impact on the demand for automobiles, utility vehicles, and trucks are the COP/USD exchange rate and the per capita income.
- **The demand for used vehicles positively impacts sales of new vehicles**: Contrary to what one might think, it was shown that an increase in the demand for used vehicles brings with it an increase in the demand for new vehicles
- **The market does not react immediately to changes in the economic context**: The demand is associated with the economic context presented in previous periods. The effects of the macro-economic variables' changes will take from 1 to 6 months to reflect a change in demand for vehicles.

REFERENCES

BBVA Research. 2016. "Situación Automotriz." Unidad de Colombia. www.bbvaresearch.com/wp-content/uploads/2016/04/SituacionAutos2016VersionFinal.pdf.

Borshchev, A. and A.Filippov. 2004. "From Systems Dynamics and Discrete Event to Practical Agent-Based Modeling: Reasons, techniques, tools." The 22nd International Conference of the System Dynamics Society, July 25–29, Oxford.

Granger, C. 1969. "Investigating causal relations by econometric models and cross spectral methods." *Econometrica*, 37: 424–438.

Mosterman, P. 1999. "An overview of hybrid simulation phenomena and their support by simulation packages." *Proceedings of the Second International Workshop of Hybrid Systems*, 165–177.

Pearson, K. 1895. "Notes on regression and inheritance in the case of two parents." *Proceedings of the Royal Society of London,* 58: 240–242.

Powell, J., N. Mustafee, S. Brailsford, et al. 2015. "Hybrid simulation studies and hybrid simulation systems: definitions, challenges, and benefits." *Proceedings of the 2015 Winter Simulation Conference*, 1678–1692.

Rogers, E. 1983. *Diffusion of Innovations.* 3rd Edition. New York: The Free Press.

Shanthikumar, J. and R. Sargent. 1983. "A unifying view of hybrid simulation/analytic models and modeling." *Operations Research,* 31, 1030–1051.

Schiffman, L. and J. Kanuk. 2014. *Consumer Behaviour.* Global Edition. New York: Pearson Education Limited.

Sterman, J. 2000. *Business Dynamics: Systems thinking and modeling for a complex world.* New York: McGraw-Hill.

Swinerd, C. and K. McNaught. 2011. "Design classes for hybrid simulations involving agent-based and system dynamics models." Simulation Modelling Practice and Theory, 25: 118–133.

7 Beyond the Seaport

Assessing the Impact of Policies and Investments on the Transport Chain

Mamoun Toukan, Hoi Ling Chan,
Christopher Mejía-Argueta, and
Nima Kazemi

7.1 INTRODUCTION

Seaports play an instrumental role in enabling global trade (Yeo, Thai, and Roh 2015). Maritime transport accounted for 80% of the US$15.46 trillion annual trade value in 2017, of which global containerized trade usually represents approximately 70% of the total (UNCTAD 2017). To stay competitive, seaports now seek to offer strategic integration into the inland supply chain. This requires interactions among container terminals, customs, inspection authorities, ship agents, freight forwarders, carriers, etc. (Lam and Song 2013). However, all of these interactions add to the logistics complexity, given that an improvement in one process may be a constraint to another (Wong 2017).

Failing to consider the most significant interdependencies might result in sub-optimal solutions (Ng and Lam 2011), or cause problems, such as the worst congestion case of Aqaba terminal in 2003, which forced ships to wait 150 hours to berth (Cebotari and Dennis 2008). Though Jordanian ports and their supply chains have improved significantly since 2003's congestion, Jordan ranked 84 in the Logistics Performance Index (LPI) in 2018 (World Bank 2018a). As an import-dependent country, with a growing population rate of 45% and a rise in the gross domestic product (GDP) by approximately 76% between 2008 to 2016, Jordan is expected to experience higher trade volumes in the years to come (see Figure 7A.1 in Appendix 7A), despite its continuing dependence on Aqaba as its only gateway to the sea. Although Aqaba's supply chain is complex, continuous investment in terminal yard space will not necessarily help with the increase in container volumes, since there are already effective policies and infrastructure in place. Instead, by understanding the interactions among activities that exist in the transport chain, policymakers can assess interventions and ensure they have a positive system-wide outcome.

In this chapter, we develop a system dynamics framework that assesses the impact of investment decisions on the container transport chain and regulations

DOI: 10.1201/9781003137993-8

in Jordan's ports. Our study contributes to the industry by providing answers to the following questions: (1) What are the main activities from Aqaba Container Terminal and their effects into the supply chain?; (2) How do these activities relate to each other and affect the delivery times and the transport chain?; and (3) What are the impacts of investing in assets and changing regulations on delivery time and the transport chain?

We address the first two questions by mapping the containerized import process in Jordan via a set of interviews with local stakeholders. The interactions among stakeholders and the terminal's processes are explicitly modeled in a causal loop diagram. Finally, some trends about container movements and historical volume records are used to identify patterns. To answer the third question, we perform a simulation based on a system dynamics framework. The chapter acts as a starting point for how policymakers can make use of system dynamics to assess impacts of maritime- or transport-related policy and their potential investment strategies. Despite the scenarios being simplifications, they are stylized cases to model the reality and might be extended to create more complex analyzes.

As each city and seaport's transport chain are unique, our study does not provide general recommendations on what policies or transport-related investments policymakers should advocate; rather we show how our approach would help them assess direct and indirect impacts of multiple solutions. This study was conducted in 2017, and the case study relates to Jordan's transportation chain focusing on data from that year backward. However, we have included a brief comment on how the responses to the COVID-19 outbreak in 2020 by some governments have impacted their seaports and transportation chains. Then, we briefly link them to Jordan's experience with the COVID-19 lockdown and the study at hand, to further emphasize the study's continued relevance and importance.

The rest of the chapter presents: a brief literature review in Section 7.2; the proposed methodology in Section 7.3; an overview of the case study on Jordan's container transport chain and the problem description in Section 7.4; the system dynamics model in Section 7.5 to test alternative investments and regulations under multiple scenarios; and finally, conclusions are drawn in Section 7.6.

7.2 LITERATURE REVIEW

With the growth of containerized trade, shipping lines capacity is also increasing, thereby requiring container terminals to be accommodating. A container terminal is an area where containers are placed prior to being exported to other countries or imported to the mainland. Maritime container terminals are key nodes for global supply chains, as buffer zones between the modes of transport, and to absorb the delays created in other parts of the supply chain (Kourounioti, Polydoropoulou, and Tsiklidis 2016). Thus, the industry is currently following two main trends: expanding capacity and increasing integration to hinterlands. A terminal's productivity does not only depend on its operational efficiency but also on its stakeholders'. In the Korean port system, for example, the underlying issues relate to external parties, like government policy (Ha, Yang, and Heo 2017).

7.2.1 INTERNATIONAL CONTAINER TRANSPORTATION

To achieve greater economies of scale, ocean carriers are heavily investing and deploying megaships, with capacities greater than 8,000 twenty-foot equivalent units (TEUs) (Prokopowicz and Berg-Andreassen 2016). With larger volumes of containers transported, ports are under tremendous pressure to discharge and load containers in a short period to avoid delays and congestion. The rise of containerized trade and lack of capacity at the terminals are increasing bottlenecks in the landside transport system serving the seaports (Roso and Lumsden 2009).

7.2.2 POLICYMAKING FOR SEAPORTS

To accommodate the surge in container volumes, the development of dry ports in the hinterlands became crucial (Acciaro and Mckinnon 2013). A dry port is an inland, intermodal terminal connected to seaports, where customers can drop off and collect containers as if they are directly doing it at the seaport (Roso and Lumsden 2002). The customs inspection is handled outside the seaport terminal (Roso, Woxenius, and Lumsden 2009). Advantages of dry ports include increasing effectiveness in terminal operations, higher level of service, and lower storage rates, among others (Panayides and Cullinane 2002; Roso and Lumsden 2010; United Nations ESCAP 2015).

Between 2004–2005, an unpredicted rise in the volumes of containers shipped from China caused congestion of terminals between Hamburg and Le Havre ports in Europe. Solving the issue brought the need to consider the Extended Gate Concept (EGC). Under EGC, congestion can be reduced by allowing container terminals to push blocks of containers to nearby dry ports and allowing import containers to be picked up by importers, instead of forcing long waiting times to berth (Veenstra, Zuidwijk, and Van Asperen 2012).

7.2.3 THE RESEARCH GAP AND OPPORTUNITY

Most of the studies about container transport chain have focused on the container terminal or hinterland operations (Roso 2008; Roso, Woxenius, and Lumsden 2009; Clausen and Kaffka 2016; Wiegmans and Witte 2017; Qiu and Lee 2019; Sislioglu, Celik, and Ozkaynak 2019). Practitioners and academicians did not take a holistic approach to identify how policy changes may affect the container delivery time and the transport chain as a whole system. This study aims to investigate impacts of the change in policies on the container transport chain using system dynamics.

In addition, we did not find investigations about the Jordanian container transport system, which is the focus of this study. Thus, this study contributes to the literature by developing a systematic approach for Jordan's containers transport chain, which aims to analyze the impact of policy changes and investments in the whole transport chain. The study is generalizable to other developing countries, as the developed framework can be easily applied to similar countries with relatively poor infrastructures and low LPI. The study also provides opportunities to assess and improve policymaking linked directly or indirectly to port performance.

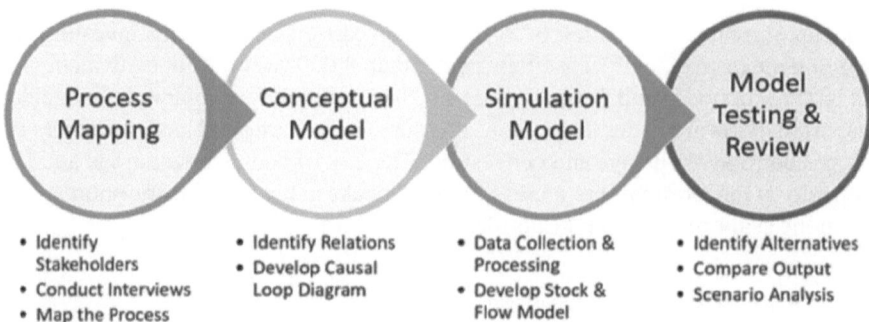

FIGURE 7.1   Overview of the methodology

7.3   METHODOLOGY

This section outlines the procedure that we follow to answer the research questions and create a framework that helps decision makers model the impacts of external decisions on the seaport's performance. Figure 7.1 shows the methodology followed to create a model that simulates Jordan's containerized transport chain at Aqaba Container Terminal (ACT). We focus on the containerized import process.

7.3.1   Process and Stakeholder's Mapping

We collected primary data based on semi-structured interviews carried out with ACT staff and other stakeholders such as local freight forwarders, ship agents, transporters, and custom brokers. Each stakeholder was asked about their role in the transport chain, the process for which they are responsible, and their key performance indicators (KPIs). We focused on gathering data from three supply chain flows: physical, information, and cash (see Appendix 7A). The interactions with other stakeholders and the complexity that exist in the container transport system were also collected in order to understand the operational system (see Figure 7A.2 in Appendix 7A).

7.3.2   Secondary Data Collection

We collected data from ACT and Jordan's Trade & Transport Commission to estimate capacities and annual volumes. Due to the lack of available data centers for processing times and transport times, semi-structured interviews with local industrial stakeholders were conducted to estimate the delays. The data collected served as bottom-line for our proposed simulation model. The study was conducted in 2018, and hence used data from prior years, mostly 2016 and 2017. We have included a brief overview of Jordan's transportation chain experience during the COVID-19 lockdown in 2020, as it emphasized the importance of the study.

7.3.3 System Dynamics Model

System dynamics is a method for studying and managing complex feedback systems to gain insights about inner structure and behaviors (Forrester 2016). Thus, system dynamics focuses on the interactions among components of a system (Sterman 2000), allowing stakeholders to design policies that seek to eliminate undesired patterns (Kirkwood 1998). Moreover, system dynamics has proved to be an appropriate approach to study a complex maritime transportation system (Oztanriseven et al. 2014). Authors have modeled freight rates evolution in global shipping (Veenstra and Lidema 2003), port operation systems based on time, quality, and profit (Mei and Xin 2010), and port performance metrics (Ridwan and Noche 2018).

After mapping processes, a causal loop diagram (CLD) is constructed to illustrate the interactions that exist between the stakeholders and processes. The core subsystems are identified based on the process mapping along with the variables that affect supply chain flows within each sub-system. Variables are connected with an arrow showing a positive or negative polarity. The CLD focuses on analyzing how changes among variables affect KPIs like time delays and terminal's performance in the long term. Based on the CLD, we create a stock-and-flow model to simulate the decision-making process and derive numeric outcomes based on behaviors and patterns, which provide practical insights.

7.3.4 Model Validation

We tested our methodology using the Status Quo (i.e., *business as is* case) and three transport strategy alternatives. These alternatives were: (1) increasing capacity in the hinterland, i.e., investment in a dry port; (2) reducing documentation time, i.e., by improvements in technology integration between customs and other parties; and (3) combining the previous strategies. We built these strategies based on industry trends and initiatives that the Jordan government will make over the coming years, according to interviews.

To evaluate these strategies, four KPIs were proposed and condensed into an index. The best output performance was ranked with a value equal to four, and the lowest with a value equal to one, and the KPIs were given equal weights. The KPIs considered for this study were: (1) container turnaround (i.e., the number of days for a container to leave and return to the terminal); (2) truck utilization (i.e., the number of trips required to complete a container turnaround); (3) delivery time (i.e., the number of days to deliver a container); and (4) container acceptance—indicates yard congestion, which reveals whether all containers on board a ship or returning back to the terminal (empty or for export) can be accepted. Container acceptance is similar to yard utilization, which is measured in percentage: we have used a binary notation to simplify the function. Yard utilization would impact time to move a container within the terminal, and a utilization of a 100% would mean the terminal can no longer accept containers. If the terminal's yard gets congested, then containers would be rejected from being discharged or gated in.

The three strategies were also tested under three scenarios and ranked. The scenarios included: (1) constraints in terminal yard capacity; (2) trucking capacity;

and (3) documentation capacity. The constraints surged due to changes in demand or supply shocks. The simulation tests were run with: (1) a 30-day horizon with one-ship arrival to highlight the turnaround cycle; and (2) a one-year horizon with a daily ship arrival.

7.4 CASE STUDY: JORDAN'S CONTAINER TRANSPORT CHAIN

7.4.1 PROBLEM DESCRIPTION

Located in the Middle East, Jordan has a population of around nine million, and accounted for a GDP of US$40.68 billion in 2017 (World Bank 2018b). In 2018, Jordan ranked 84 in the LPI, down from 67 in 2016 (World Bank 2018a). Showing a privileged location in the northern Red Sea, at the junction of trading routes, Aqaba has not only been Jordan's unique access to the sea but also it has emerged as the third largest Red Sea port, after Suez in Egypt and Jeddah in Saudi Arabia (UNDP 2012).

Aqaba's terminal experienced the worst congestion in mid-2003, increasing anchorage-waiting times to 150 hours for vessels docking at the port (Cebotari and Dennis 2008). In consequence, Aqaba Development Corporation (ADC) signed a two-year contract with APM terminals in March 2004 to avoid facing similar problems in the future. Based on APM's performance, a joint venture agreement with ACT was announced in 2006 (Cebotari and Dennis 2008) to continue achieving a proper performance at Aqaba's terminal. Aqaba also served as a gateway to Iraq, but due to political escalations in Iraq the border was shut in 2014. The re-opening of the Iraqi border may raise the volume of cargos transited through Aqaba's terminal (Kayed 2018).

Thus, to ensure Aqaba keeps competitive for domestic and transit cargo, policymakers must ensure a forward-looking, adequate mix of policy and investment strategies to ensure continuing efficiency of ACT and the containerized transport chain as a whole. This will allow ACT to reduce its 2016 average import dwell time of nine days (the number of days that importing containers need to wait to gate out from the terminal). Port and terminal authorities can modify the container dwell time to gain space and increase the capacity of storage yards. Studying large port terminals, Merk (2013) reports: around 110,000 TEUs handled per crane; 25–40 crane moves per hour; and a dwell time for import boxes of 5–7 days, and for export boxes 3–5 days. According to a Jordanian newspaper, the Jordanian government has a target of reducing the dwell time to three days through improvements in customs clearance and infrastructure investments (JT 2017).

7.4.2 MAPPING THE PROCESS

The import process depends on flows of documentation, information, cash, and physical flows. The documentation flow was separated from the information flow as it has a physical flow of the documents involved, which is different from cargo movement. Figure 7A.2 in Appendix 7A summarizes the flow of imported containers in Jordan, including the parties involved. According to ACT's data, dwell time (the number of days it takes to clear an import container from the terminal) was eight days, with an additional 1–2 days to deliver the container to the final destination and another 1–2 days to return the container. A significant portion of the time to deliver an imported container to the final destination is allocated to activities outside the terminal's gates. The involved parties located in Amman, where head offices of the shipping lines and the majority of cargo receivers are based; ACT; and Aqaba Customs Yard 4, where inspection takes place. According to interviews conducted, these distances/times represent a significant factor that contributes to the time to deliver an import container from ACT.

7.4.3 THE CONCEPTUAL FRAMEWORK

Figure 7.2 provides a holistic view of Jordan's container transport chain (i.e., the system), its elements (i.e., sub-systems), and how they interact with one another. For the sake of brevity, we combined certain elements, such as the documentation processing time, the ship agent, and customs procedures under the same sub-system of the CLD. Certain time variables such as customs inspection time and container's pick-up time were skipped to avoid redundancies, as they are affected by the dwell time and affect the delivery time. External factors, which are beyond the system's control, like macro-economic and political factors, have been excluded.

Based on the feedback loops (see also Table 7B.1 in Appendix 7B for further details), ACT is constrained by its yard capacity and its equipment. The documentation process is constrained by the number of agents, customs brokers, and inspection personnel. Additionally, the trucking system is constrained by the number of vehicles, drivers, and the number of allowed monthly trips. Terminal's performance is not only affected by the internal operations but also by external factors. Hence, there will be a limit to how much improvement may be achieved from internal solutions within the terminal.

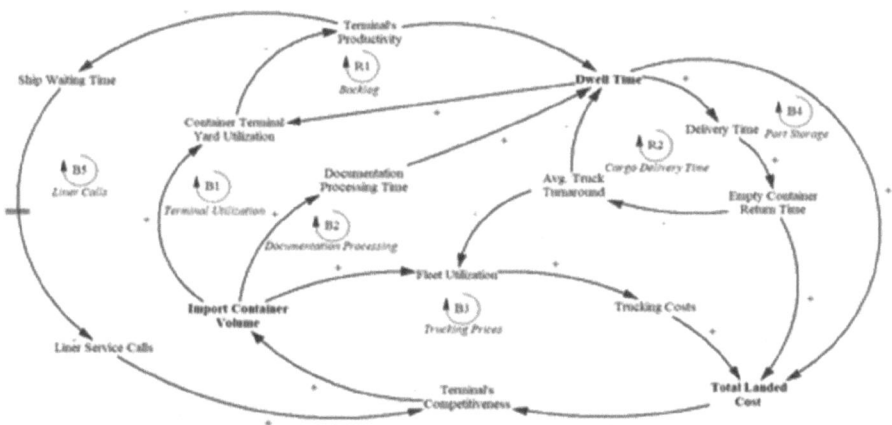

FIGURE 7.2   CLD model of Jordan's container transport chain

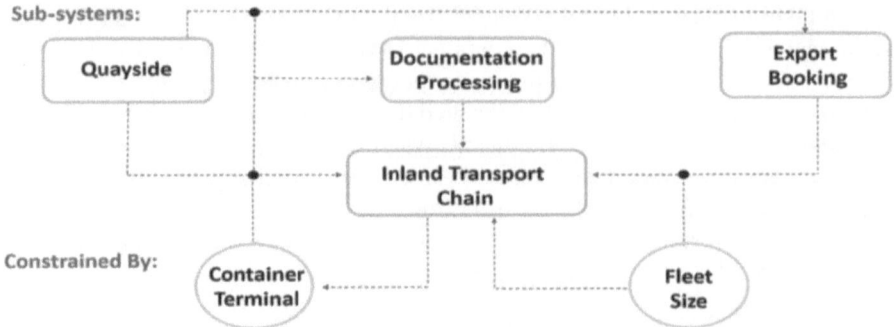

FIGURE 7.3   Core sub-systems and main constraints

7.4.4   Driving System Dynamics into Practice: A Simulation Approach

Building upon the CLD, the container transport chain is broken down into the following sub-systems: (1) quayside sub-system, (2) documentation processing, (3) import container movement, and (4) export container movement. Figure 7.3 illustrates the sub-systems and constraints that we further develop in this study. A limitation of our study is that we consider containers as a generic unit, without differentiating types or sizes; truck sizes are homogenized by loading only one container per truck. In addition, time is defined in days and there is no initial inventory.

For further details about assumptions and the stock-and-flow model, see Appendix 7C.

7.5   DISCUSSION AND ANALYSIS OF RESULTS

In this section, we present simulation outputs using the proposed methodology. Due to the lack of a unified transport databank in Jordan, and the confidentiality of non-public data, the variable inputs used are proxy numbers. However, the output mimics reality on a high level. We run two simulations: (1) a short-term simulation, with a 30-day time frame, and only one ship to arrive in order to show the flow of one batch of containers that have entered the system; and (2) a simulation that provides added insights into how the alternative strategies will perform in the long term (i.e., one year, and a daily ship arrival rate) and while relaxing some constraints. Figure 7.4 summarizes the inputs used in the model. Stock-and-flow models (see Appendix 7D and Figure 7D.2) were developed and run using Vensim, Windows Version 7.2.

The identified alternative strategies were based on interviews with local stakeholders about the expected initiatives that are in line with the global industry and can be taken to improve the logistics performance of Jordan. These strategies mainly require changes in two input variables: "open dry port" and "documentation processing time". Figure 7D.2 in Appendix 7D shows an extension to the inland container transport chain if a dry port was open. The dry port would act as a binary variable: when activated, imported containers will be shipped there. We will be referring to alternative policy 1 (i.e., moving to the hinterlands) as "Dry Port;" alternative

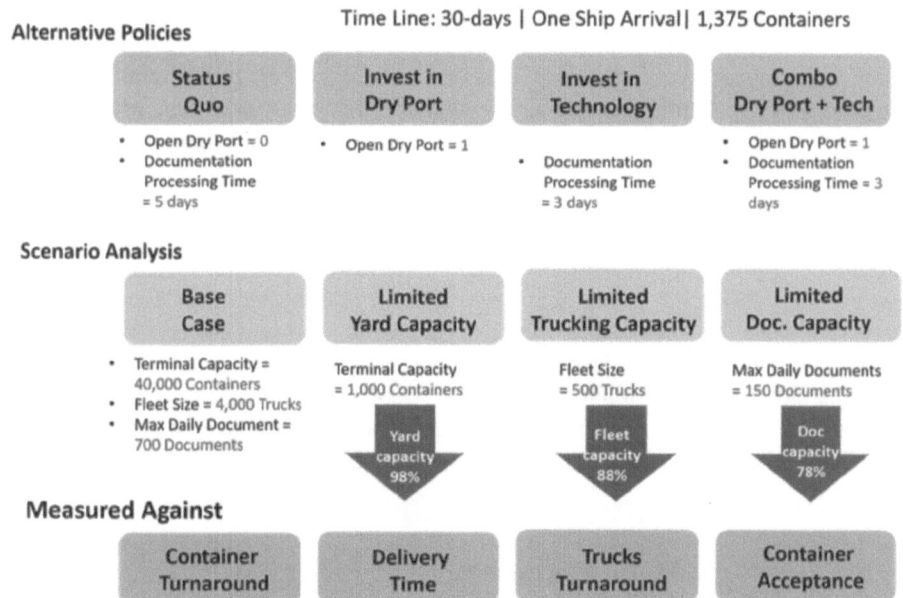

FIGURE 7.4   Summary of the scenarios and instances

policy 2 (i.e., reducing documentation time) as "Tech Investment;" alternative policy 3 as "Combo;" and the base case as "Status Quo." To facilitate the comparison, the highest-ranked alternative is chosen with the base case to run the second part of the simulations, where a longer time period was used (one year, and a daily ship arrival rate).

7.5.1   STATUS QUO

For the first part of the simulation, we assumed the main container's terminal operates for a period of 30 days, receives only one ship, no hinterland terminals/dry ports are available, and it takes on average five days to process documents. Simulating with the arrival of only one ship makes it easier to track the location of containers across the transport chain. Refer to Figure 7D.1 in Appendix 7D to see Status Quo input variables. Under the base case scenarios, it takes one day to discharge the containers from a ship at the terminal, and it takes an average of 19 days from discharge date to have all containers returned back, with the first batch of containers returning back on day 8 after the discharge date.

7.5.2   RESULTS OF STATUS QUO

Under the base scenario, the Combo policy—which includes moving containers to a dry port (at a hinterland location) and improving on the documentation processing time through investing in technology and staff training—resulted in the most preferential output, when measured against the four KPIs: container turnaround, delivery time, truck's turnaround, and container acceptance. The Tech Investment policy was ranked second. We highlight that in the Dry Port policy, containers are moved out of the terminal immediately after the discharge date, i.e., reducing the dwell time in the port, but it does not reduce the delivery time compared to the Status Quo. Hence, the real reduction in delivery time came from improving documentation processing time.

To assess the risk and resilience, each alternative policy is then tested under extreme conditions for the same list of KPIs. We introduce three scenarios: limited yard capacity, limitation in national fleet capacity, and limitation in documentation processing capacity. These three scenarios help us to represent either supply shocks or a surge in demand which leads to capacity constraints. Examples of supply shock would include strikes, national disasters, and depreciation of trucks. By testing the policies against extreme conditions and understanding the degree of risk/resilience each policy alternative has, policymakers will be able to avoid massive disruptions in the transport chain.

7.5.3 Results for Multiple Scenarios

Under the first scenario, the yard capacity was dropped to around 1,000 containers. Due to the terminal's limited capacity in Scenario 1, the terminal rejected some containers to be discharged from the ship, and more so in the alternatives that did not have a dry port (see Figure 7D.3 in Appendix 7D). Thus, the stock of containers that are returned empty or for exports do not converge to the same level, indicating that the terminal yard reaches capacity. Container acceptance is vital to ensure the terminal can accommodate the expected future surges in demand. Under this scenario, truck utilization and delivery times are not affected.

However, with daily ships arriving, the situation would be worse, as longer queues would be expected at the terminal's gate for gating-in trucks returning empty and export containers. If the container terminal yard is expected to be a constraint due to surges in demand or potential shocks within the terminal, then having access to a dry port would reduce the risk of congestion at the port. Under Scenario 1, the Combo policy had the highest outcome, being able to accommodate more containers compared to the Tech Investment policy and the Status Quo alternatives. The Combo policy achieved a lower container turnaround and delivery time compared to the alternative with a Dry Port only, as output results show in Figure 7D.3 in Appendix 7D.

Under Scenario 2, the national fleet size was dropped to 500 trucks, representing a drop of about 88% compared to the base scenario. Figure 7D.4 in Appendix 7D shows the output under Scenario 2, where the Tech Investment policy had the highest ranking. For alternatives with a Dry Port, despite imported containers being moved out from the terminal at a faster rate (i.e., a lower dwell time), the movement from the dry port to final destination takes longer compared to the Status Quo and Tech Investment policies.

Moving an imported container to a dry port prior to delivery requires double inland-transport movements—one from terminal to dry port and another from dry port to the final destination—and returning empty, or export back to the main terminal. Hence,

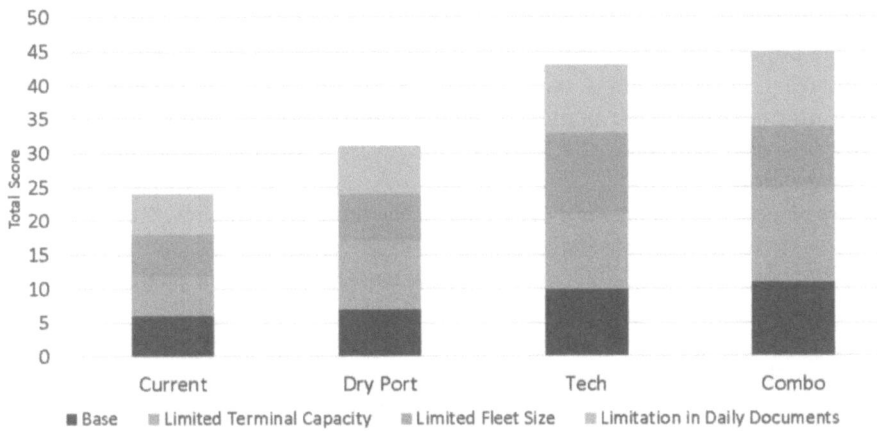

FIGURE 7.5   Overall ranking for each strategy under different scenarios

the alternative policies with a Dry Port increase truck utilization, indicating that policymakers must consider the transportation capacity when utilizing the hinterlands.

For Scenario 3, the number of documents processed for imported containers is reduced from 700 daily documents to 150 documents. We assumed each documentation belongs to one container. The reduction in documentation processing could be a result of a strike at customs, or disruption of shipping agencies, or a surge in container levels creating backlogs in processing times. Under this scenario, the containers in the terminal yard, and delivery times graph curves shifted to the right, resulting in longer delays. Again, the Dry Port and Combo policies showed higher truck's utilization and achieved lower dwell times, but the Combo alternative achieved faster delivery times and was ranked higher (see Figure 7.D5 in Appendix 7D).

We ranked each policy/strategy by giving them a score and summing the total rank obtained at each KPI for all scenarios (see Figure 7.5). Strategies with Dry Ports reduced the dwell time to four days, with most containers gating-out between one and two days. However, the delivery time was only reduced when improving the documentation's processing time. Regarding delivery time, the Tech Investment policy outperforms the Dry Port strategy. If capacity constraints are binding, Dry Port would outperform Tech Investment, as it can accommodate more containers when the container terminal reaches capacity. The Combo policy received the highest ranking and added a greater resilience to the system in the short-term simulation with the arrival of one ship only.

7.5.5   SIMULATION FOR A ONE-YEAR PERIOD

Although the Combo policy achieves lower dwell times and delivery times, it puts high pressure on the trucking industry because of the double container movement. As a result, the Combo alternative produces a higher accumulation of containers at the terminal's yard over time, i.e., increasing yard utilization (see Figure 7D.6 in Appendix 7D). At day 275 of the simulation, the terminal experiences congestion and trucks with export or empty containers can no longer gate-in, and ships can no longer discharge imports. As stated earlier, a ship needs to discharge its load prior to being able to take export and empty containers out of the terminal. Thus, the congestion in the yard translates to the berth and the transportation fleet. This will cripple the country's supply chain and entails severe congestion, possibly worse than Aqaba's 2003 congestion.

However, by increasing the fleet capacity for the Combo policy, from 4,000 to 5,000 trucks, there is a substantial reduction in the terminal's yard utilization (see Figure 7D.7 in Appendix 7D). The levels of containers in stock at the terminal yard oscillates at a significantly lower range compared to terminal capacity. The adjusted strategy (or Combo+) creates a better overall transport chain, with lower delivery time and container turnaround cycle compared to the Status Quo.

7.5.6 Managerial Insights and Potential Policymaking

This chapter highlights the importance of taking a holistic approach prior to implementing policies that might affect the whole supply chain at ports, by looking beyond the terminal gates and understanding the interactions of all the elements in the transport chain. We underline that policymakers must choose the right metrics to analyze the system depending on their bottlenecks, processes, stakeholders, constraints, and conditions. For example, a reduction in dwell time does not mean a reduction in delivery time: we showed that the Dry Port policy did result in a lower dwell time but had a similar delivery time to the Status Quo policy. Hence, the dwell time as a metric for policymakers should not be emphasized too much, compared to using delivery times or container turnaround times.

In addition, building multiple tailor-made scenarios with multi-disciplinary groups will help decision and policymakers assess various strategies under extreme conditions. In our case, reducing the documentation processing time, through technology investments, resulted in faster dwell time and delivery time. When the capacity of a terminal's yard is an issue, having faster documentation processing time alone will not solve the issue. However, leveraging both the hinterlands and technology provides the container terminal with a greater capacity and achieves lower delivery time (i.e., creates greater resilience and agility).

The use of the hinterland adds pressure on trucking, as it requires a larger number of trips. When trucking trips increase, if fleet size is not adjusted, fleet utilization increases as a result, which will eventually lead to a backlog of containers at the terminal. Thus, improving one element of the logistics chain does not lead to an overall improvement. As seen when technology and dry port investments were made in the Combo policy, the overall logistics chain was worse in the long term. Under the assumption that the same pool of trucks is used to transport containers from the terminal to the dry port and then to the destination, fleet utilization increases significantly.

The proposed framework still proves relevant and is useful today. In 2020, the world came to a standstill with the spread of the COVID-19 crisis caused by the SARS-COV2 virus. Governments around the world forced lockdowns and curfews to limit the spread of the virus. These lockdowns massively disrupted global supply chains. Factories across Asia were affected in the first half of 2020, worsening the global supply, while the reduction in social activities led to lower consumption, thus impacting demand worldwide. Several ports from around the world experienced congestions or, in other cases, decreased their operations to the lowest capacity. This ultimately led to blank sailings, disrupting supply chains even more. In Manila, for example, as only 40% of what was discharged in March 2020 was moved out of the ports' gates, the head of the Philippine Ports Authority issued a warning to inform that the port may shut down in six to eight days (ABS-CBN 2020). JNPT, Mundra Port also suffered from congestion in April 2020 because importers were not clearing containers with sufficient efficiency (Manoj 2020).

Jordan took important measures mid-March 2020, imposing a strict curfew. The government extended the number of free days at the port from 6 to 14 days and it was extremely selective at what businesses were granted permits to operate. Even as the terminal was exempted from the lockdown, other major stakeholders were not operating; thus significantly reducing the flow of containers out of the gates. The number of containers gated out were reduced by about 50%, and on some Fridays, typically a congested day at the port, the number of containers gated out was zero. The port's yard utilization got close to 80%, which is beyond the safety operating capacity. This experience showed how a disruption in any of the stakeholders will impact the port's yard utilization, and as it increases and the yard gets congested, port productivity is impacted. The terminal was fortunate to alleviate the congestion, due to the government revising certain measures periodically, such as allowing the port community that includes truck drivers and customs clearance personal to work on Friday, reducing the idle time back to six days, and granting more permits for importers to be able to unload containers.

Utilizing the presented framework would help policymakers to immediately assess the impact of the responses, to ensure they monitor the right KPIs to adapt their strategies accordingly, and to mitigate greater losses. As disruption in a port's activities, especially for a country that is highly depended on its only container terminal, would create greater issues along the supply chain, and across diverse roles at multiple levels of decision making.

7.6   CONCLUSION AND FUTURE RESEARCH

By taking Jordan as an example, this study developed a system dynamics model to assess policy and investment decisions related to the container transport chain. The developed model provides a simulation basis built upon the system dynamics' stock-and-flow approach to test different alternatives and strategies against several scenarios and KPIs. The developed framework is significantly relevant to Jordan and other emerging markets that are highly dependent on their seaports. We aim to advocate taking a holistic view when assessing impacts of different decisions due to the inherent complexities that exists in the transport chain.

This chapter focuses on giving insights to policymakers in public sectors, but private businesses and investors can use it to further assess transportation-related investment opportunities. For example, a terminal operator interested in investing in building a dry port will understand the potential implications on the underlying system if transportation capacities do not increase or certain customs policies do not change. Additionally, the developed framework can also be used to evaluate the container transport chain's level of resilience and ability to accommodate shocks in the system, like surges in container volumes.

This study can be extended in several directions. First and foremost, cost and other KPIs should be considered in addition to time. To remove this limitation, the model can be further extended to include financial implications of the alternatives on the overall system, or a specific stakeholder. Secondly, assumptions of the model, like container and truck sizes and types, can be relaxed to take into account other complexities. Third, the location of the dry port could be assessed by comparing the investment value and implications on the total landed cost. Finally, other alternative decisions such a rail option can be considered in the analysis. A rail option could connect the terminal to the dry port, and thus, reduce pressure on trucking. This study presents future opportunities to be explored through cooperation between industry practitioners, policymakers, and academics.

REFERENCES

ABS-CBN, N. 2020, March 31. "Manila Port May Shut Down in 6 Days as it Gets Swamped by Undelivered Cargo: Port authority." https://news.abs-cbn.com/business/03/31/20/manila-port-may-shut-down-in-6-days-as-it-gets-swamped-by-undelivered-cargo-port-authority (accessed December 19, 2020).

Acciaro, M. and A. Mckinnon. 2013. "Efficient hinterland transport infrastructure and services for large container ports." *International Forum Discussion Papers*, 2013/19.

Cebotari, D. and A. Dennis. 2008. "Jordan: A public-private partnership brings order to Aqaba's port," in *Celebrating Reform 2008: Doing Business Case Studies*. Washington DC: World Bank: 72–78.

Clausen, U. and J. Kaffka. 2016. "Development of priority rules for handlings in inland port container terminals with simulation." *Journal of Simulation*, 10 (2): 95–102.

Forrester, J. W. 2016. "Learning through system dynamics as preparation for the 21st Century." *System Dynamics Review*, 32 (3–4): 187–203.

Ha, M. H., Z. Yang, and M. W. Heo. 2017. "A new hybrid decision making framework for prioritising port performance improvement strategies." *Asian Journal of Shipping and Logistics*, 33 (3): 105–116.

JT. 2017. "ASEZA boosts Aqaba Port capabilities by minimising container clearance time." *The Jordan Times*. www.jordantimes.com/news/local/aseza-boosts-aqaba-port-capabilities-minimising-container-clearance-time.

Kayed, M. 2018. "Truck movement between Jordan, Iraq to return to normal in March." *The Jordan Times*. www.jordantimes.com/news/local/truck-movement-between-jordan-iraq-return-normal-march.

Kirkwood, C. W. 1998. "System Dynamics Methods: A quick introduction." Arizona State University. www.public.asu.edu/~kirkwood/sysdyn/SDIntro/preface.pdf.

Kourounioti, I., A. Polydoropoulou, and C. Tsiklidis. 2016. "Development of models predicting dwell time of import containers in port container terminals – An artificial neural networks application." *Transportation Research Procedia*, 14: 243–252

Lam, J. S. L. and D.-W. Song. 2013. "Seaport network performance measurement in the context of global freight supply chains." *Polish Maritime Research*, 20: 47–54.

Manoj, P. 2020, April 02. "JNPT, Mundra Port Stares at Congestion as Container Freight Stations Get Clogged." www.thehindubusinessline.com/economy/logistics/jnpt-mundra-port-stares-at-congestion-as-container-freight-stations-get-clogged/article31240703.ece (accessed December 19, 2020).

Mei, S. and H. Xin. 2010. "A system dynamics model for port operation system based on time, quality and profit." *2010 International Conference on Logistics Systems and Intelligent Management,* 3: 1669–1673.

Merk, O. 2013. "The Competitiveness of Global Port-Cities: Synthesis report." OECD Regional Development. www.oecd.org/cfe/regionaldevelopment/Competitiveness-of-Global-Port-Cities-Synthesis-Report.pdf.

Ng, T. S. and S. W. Lam. 2011. "Dynamic maritime systems inquiry: The DIVER approach." *Systems Engineering,* 14 (3): 239–254.

Oztanriseven, F., L. Perez Lespier, S. Long, et al. 2014. "A review of system dynamics in maritime transportation." *IIE Annual Conference and Expo 2014*: 2447–2456.

Panayides, P. M. and K. Cullinane. 2002. "Competitive advantage in liner shipping: A review and research agenda." *International Journal of Maritime Economics,* 4 (3): 189–209.

Prokopowicz, A. K. and J. Berg-Andreassen. 2016. "An evaluation of current trends in container shipping industry, very large container ships (VLCSs), and port capacities to accommodate TTIP increased trade." *Transportation Research Procedia,* 14: 2910–2919.

Qiu, X. and C. Y. Lee. 2019. "Quantity discount pricing for rail transport in a dry port system." *Transportation Research Part E: Logistics and Transportation Review,* 122: 563–580.

Ridwan, A. and B. Noche. 2018. "Model of the port performance metrics in ports by integration six sigma and system dynamics." *International Journal of Quality and Reliability Management,* 35 (1): 82–108.

Roso, V. 2008. "Factors influencing implementation of a dry port." *International Journal of Physical Distribution and Logistics Management,* 38 (10): 782–798.

Roso, V. and K. Lumsden. 2002. "Dry Port Concept for Seaport Inland Access with Intermodal Solutions." Gothenburg: Chalmers University of Technology.

Roso, V. and K. Lumsden. 2009. "The Dry Port Concept: Moving seaport activities inland?" *Transport and Communications Bulletin for Asia and the Pacific,* 78 (78): 87–101.

Roso, V. and K. Lumsden. 2010. "A review of dry ports." *Maritime Economics and Logistics,* 12 (2): 196–213.

Roso, V., J. Woxenius, and K. Lumsden. 2009. "The dry port concept: Connecting container seaports with the hinterland." *Journal of Transport Geography,* 17 (5): 338–345.

Sislioglu, M., M. Celik, and S. Ozkaynak. 2019. "A simulation model proposal to improve the productivity of container terminal operations through investment alternatives." *Maritime Policy and Management,* 46 (2): 157–177.

Sterman, J. 2000. *Business Dynamics: Systems thinking and modeling for a complex world.* New York: McGraw-Hill.

UNCTAD. 2017. *Review of Maritime Transport.* New York: United Nations.

UNDP. 2012. *Aqaba, Jordan, Port Expansion Case Study (Transportation).* New York: United Nations Office for South-South Cooperation: 1–7.

United Nations ESCAP. 2015. "Importance of Dry Ports in Integrating Infrastructure Networks." National Seminar on Integrated Intermodal Transport. www.unescap.org/sites/default/files/Session%203b%20Importance%20and%20benefits%20of%20Dry%20Ports.pdf.

Veenstra, A. and M. Lidema. 2003. "Cyclicality in the Oil Tanker Shipping Industry." Rotterdam School of Economics/Centre for Maritime Economics and Logistics, Latvia.

Veenstra, A., R. Zuidwijk, and E. Van Asperen. 2012. "The extended gate concept for container terminals: Expanding the notion of dry ports." *Maritime Economics and Logistics,* 1 (1): 14–32.

Wiegmans, B. and P. Witte. 2017. "Efficiency of inland waterway container terminals: Stochastic frontier and data envelopment analysis to analyze the capacity design and throughput efficiency." *Transportation Research Part A: Policy and Practice*, 106: 12–21.

Wong, A. 2017. "Port Policy: Key design issues and impact." 30th IAPH World Port Conference. www.drewry.co.uk/news/news/port-policy-key-design-issues-and-impact.

World Bank. 2018a. "Logistics Performance Index." https://lpi.worldbank.org/international/global/2018.

World Bank. 2018b. "The World Bank data: Jordan." https://data.worldbank.org/country/jordan.

Yeo, G. T., V. V. Thai, and S. Y. Roh. 2015. "An analysis of port service quality and customer satisfaction: The case of Korean container ports." *Asian Journal of Shipping and Logistics*, 31 (4): 437–447.

APPENDIX 7A: JORDAN'S CONTAINER TRADE FLOWS

Interviews were conducted in the form of open-ended questions by phone and in-person meetings. The list includes interviewees from the private sector in Jordan working in shipping, logistics, and inland transportation, as well as from the governmental sector like Jordan's Land Transport Regulatory Commission. Special thanks to Aqaba Container Terminal for providing us with insights and data.

In-transit container volumes decreased by 82% from 2014 to 2017, which represents a 91% drop from 2012, when ACT handled 104,000 TEUs in in-transit containers. The significant drop for in-transit volumes is due to the closure of the Jordanian–Iraqi borders, due to political turmoil in Iraq.

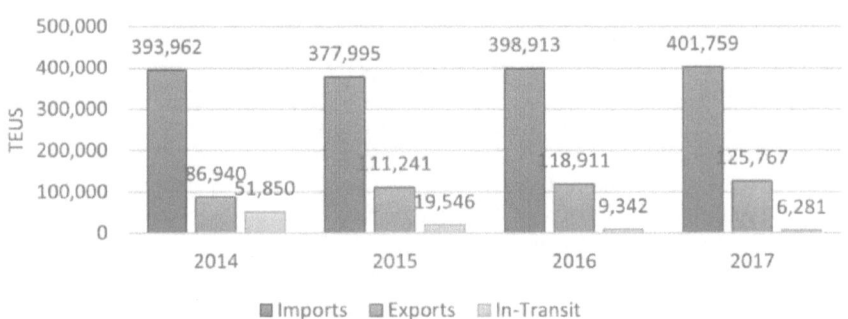

FIGURE 7A.1   Breakdown of full containers

Source: Aqaba Container Terminal.

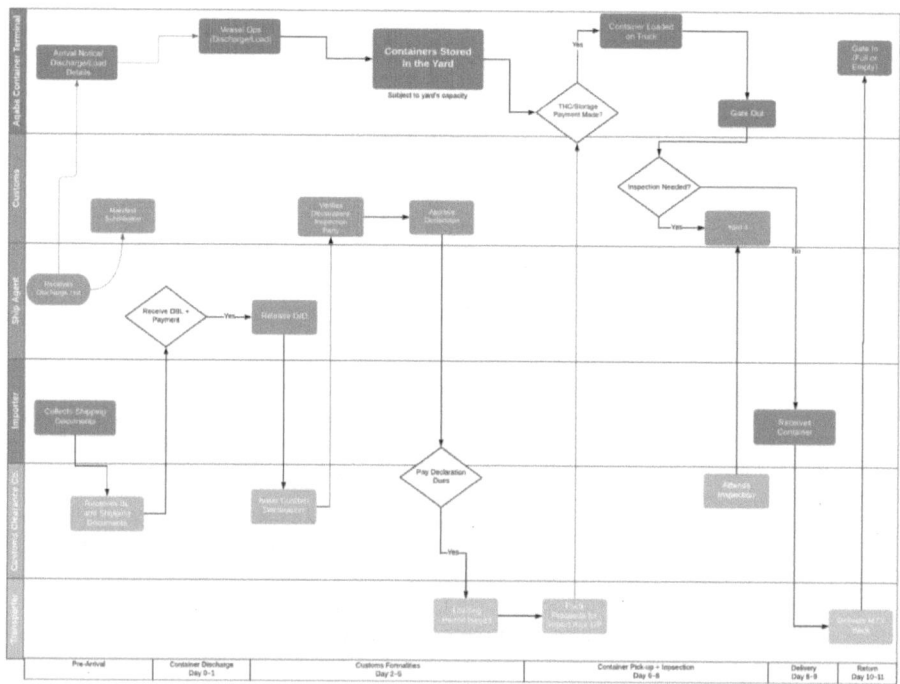

FIGURE 7A.2 Jordan's import process

APPENDIX 7B: DETAILS ABOUT THE CAUSAL LOOP DIAGRAM AND ITS VARIABLES

TABLE 7B.1
Description of Feedback Loops

Loop	Loop Name	Rising Import Container Volumes (All Else Equal)
B1	Terminal Utilization	Increase in import volumes increases the yard's occupancy, which increases the yard utilization. As containers in the yard get close to capacity, terminal productivity decreases, impacting the time it takes to service container pick-ups by trucks, leading to a higher dwell time, and ultimately increasing the total landed cost.
B2	Documentation Processing	Increase in import volumes increases the number of documents to be processed, which increases the processing time, leading to a higher dwell time, higher cost, and ultimately a higher total landed cost.
B3	Trucking Prices	Increase in import volumes requires more trucks, leading to a higher utilization of trucks. If no investments in the national fleet size is made, capacity becomes limited, leading to higher transport prices, increasing the total landed cost and reducing the terminal's competitiveness.

(continued)

TABLE 7B.1 Continued
Description of Feedback Loops

Loop	Loop Name	Rising Import Container Volumes (All Else Equal)
B4	Port Storage	The longer time it takes to pick up a container, the higher the port storage charges will be incurred, leading to a reduction in the terminal's competitiveness.
B5	Liner Calls	Reduction in the terminal's productivity means that arriving ships may need to wait longer to be serviced. This would increase the cost on the carrier, and with time carriers will either introduce surcharges, or halt certain services that call at Aqaba.
R1	Backlog	Increase in container terminal yard utilization leads to lower productivity, which leads to a higher dwell time. As containers stay longer at the terminal, the terminal's utilization increases further, ultimately leading to longer waiting times.
R2	Cargo Delivery Time	As the time to deliver containers increases, the time to return the empty container back to the terminal increases, leading to a lower turnaround time for the trucks. As truck turnaround decreases, the time to pick up a container increases, ultimately leading to a further reduction in truck turnaround and increase in delivery time.

APPENDIX 7C: STOCK-AND-FLOW MODEL

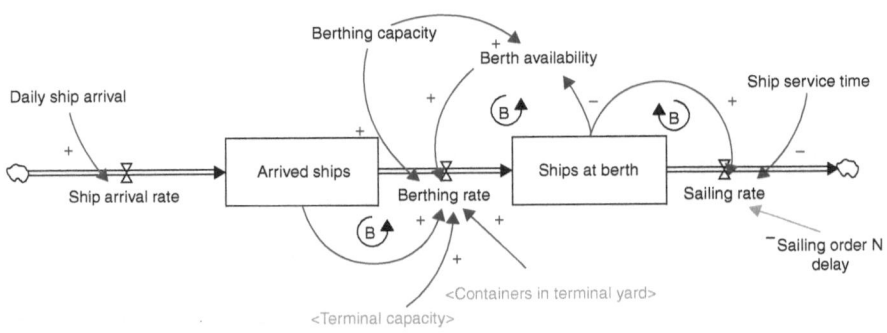

FIGURE 7C.1 Quayside sub-system

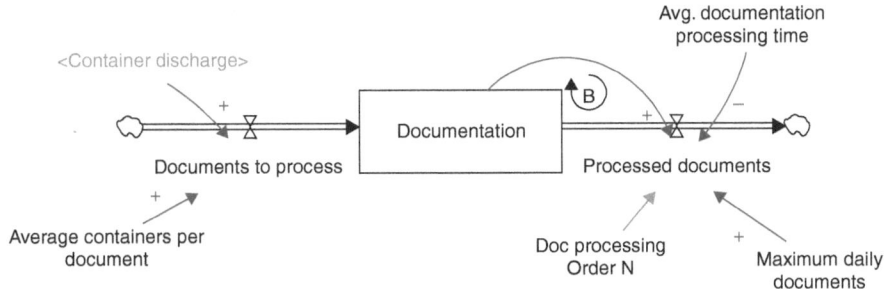

FIGURE 7C.2   Documentation sub-system

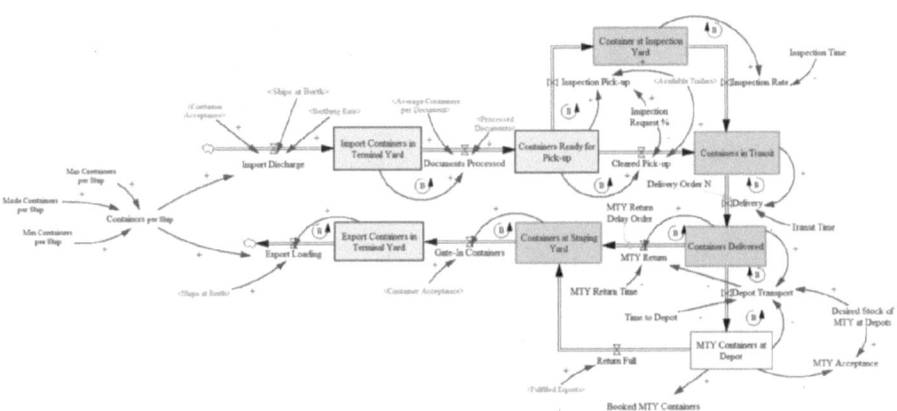

FIGURE 7C.3   Inland container transport chain

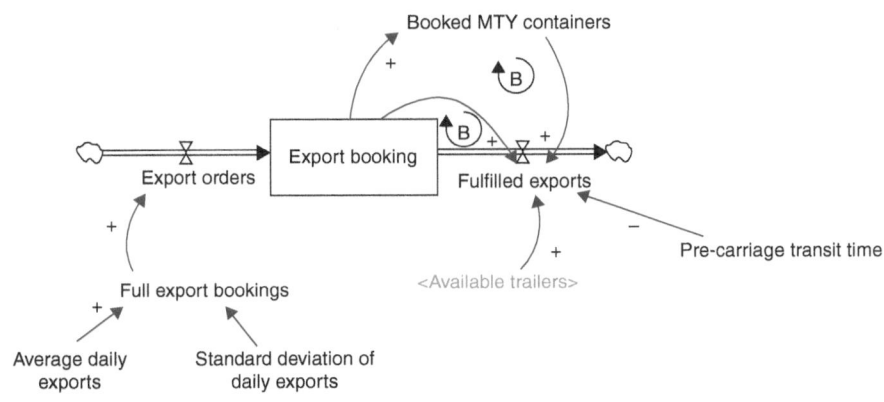

FIGURE 7C.4   Export booking sub-system

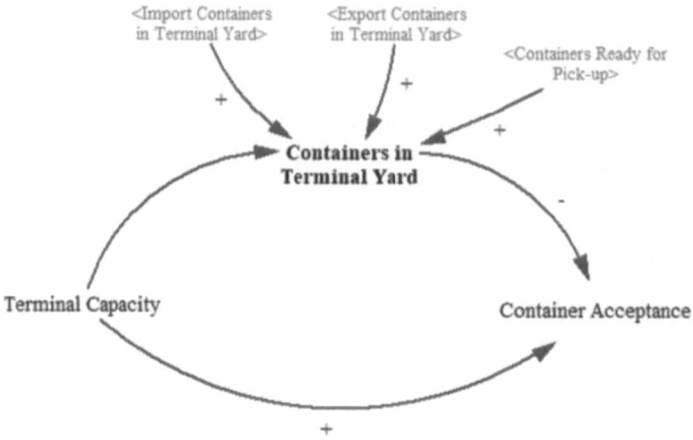

FIGURE 7C.5   Container terminal yard capacity

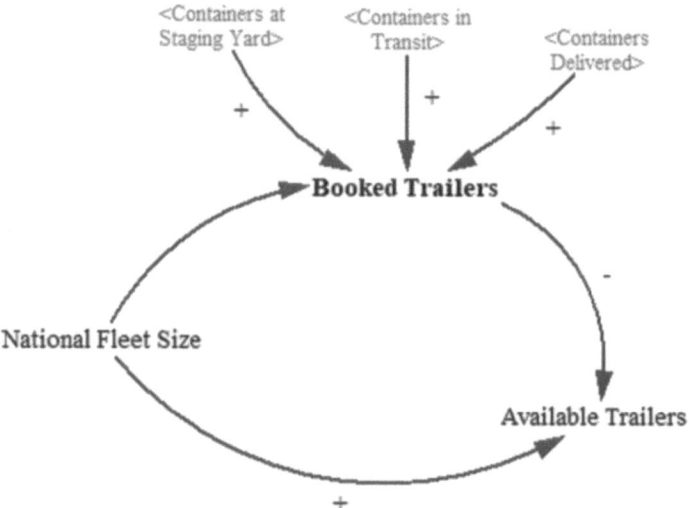

FIGURE 7C.6   Truck utilization

LIST OF ADDITIONAL ASSUMPTIONS

- Berthing rate depends on berth and container availability in the terminal yard.
- Terminal productivity affects ship transshipment rates. It serves at a constant rate of 100%, and only drops to 0% if yard utilization becomes 100%.
- Documentation process includes the documentation and the customs processes.
- Documentation process starts once containers are discharged to the terminal yard.
- Empty containers for exports are picked up from a container depot only.

- The number of empty containers that are dropped off at the depot can be pre-determined.
- When the stock of containers in the yard reaches its capacity, no container can be handled until the number of containers in stock is reduced.
- Trucks pick up containers instantaneously. No delay is caused by communications or truck's repositioning.

APPENDIX 7D: SIMULATION MODEL

STATUS QUO INPUTS

Status Quo Inputs			
Variables	**Inputs**	**Variables**	**Inputs**
Daily Ships Arrival	1 Ship	Desired Stock of MTY at Depots	500 Containers
Containers per Ship	1375 Container	Avg. Daily Exports	220 Containers
Inspection Requests	30%	Standard Deviation of Exports	50 Containers
Containers per Document	1 Container	Fleet Size	4,000 Trucks
Documentation Processing Time	5 days	Terminal Capacity	40,000 Containers
Max Daily Documents Processed	700 Documents	Open Dry Port	0 (Binary Variable)

FIGURE 7D.1 Status Quo variables

MOVING TO HINTERLANDS: EXTENDED MODEL

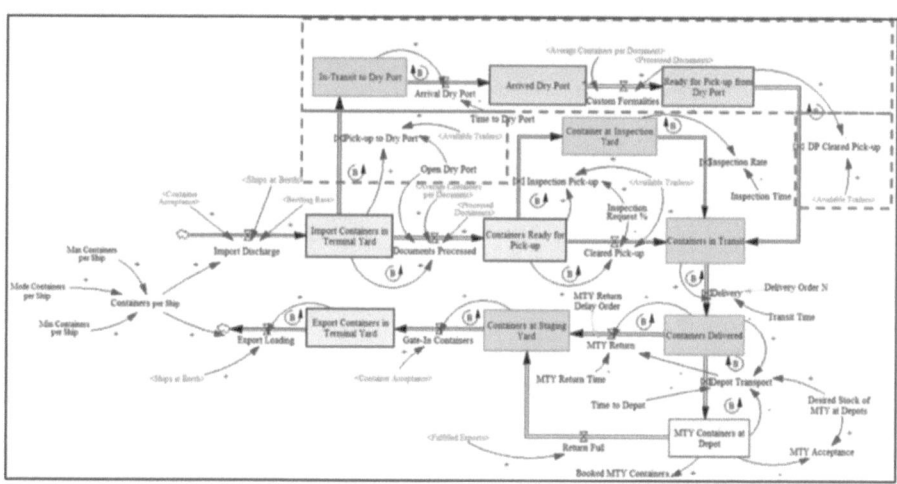

FIGURE 7D.2 Inland container transport chain

SIMULATION OUTPUTS

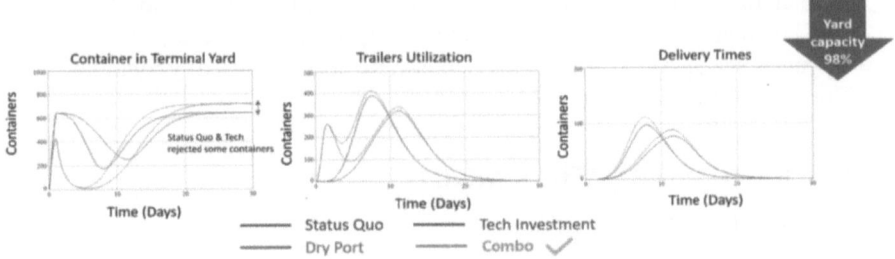

FIGURE 7D.3   Output for diverse strategies under Scenario 1

Note: A color version of this figure can be found in the ebook.

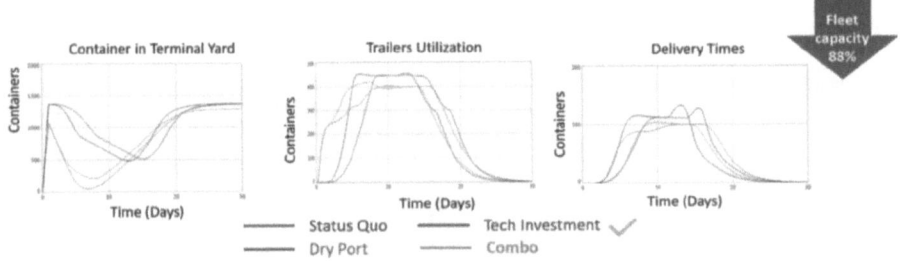

FIGURE 7D.4   Output for diverse strategies under Scenario 2

Note: A color version of this figure can be found in the ebook.

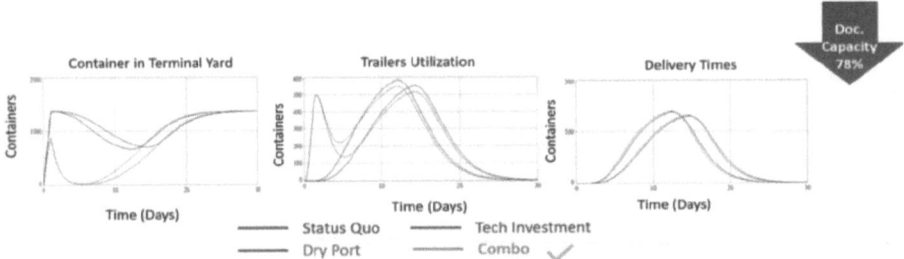

FIGURE 7D.5   Output for diverse strategies under Scenario 3

Note: A color version of this figure can be found in the ebook.

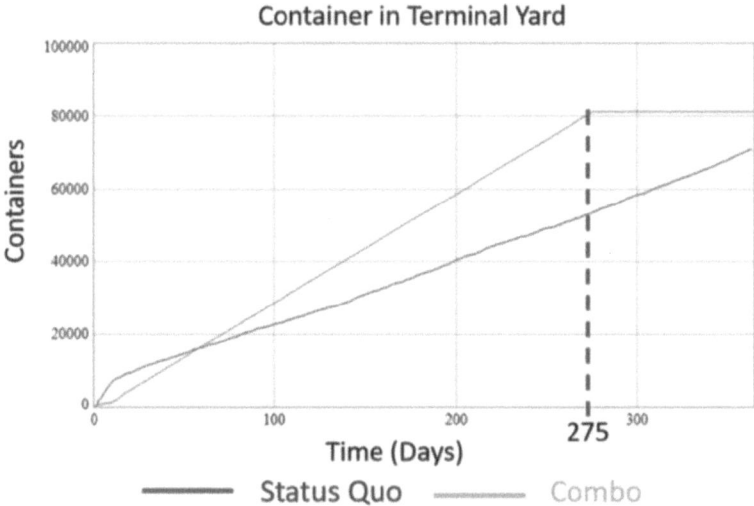

FIGURE 7D.6   Simulation output for 365 days

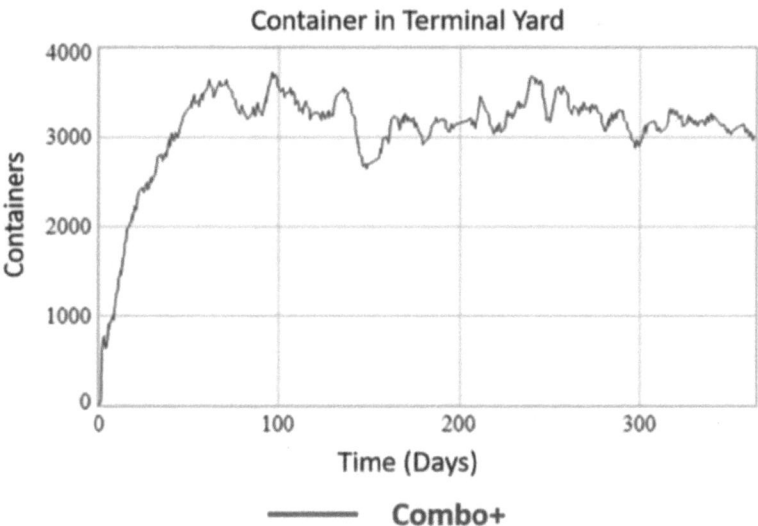

FIGURE 7D.7   Simulation output for 365 days and increasing fleet size

8 Challenges and Approaches of Data Transformation

Big Data in Pandemic Times, an Example from Colombia

Luis Carlos Manrique Ruiz and Sandra Puentes

8.1 INTRODUCTION

8.1.1 COVID-19 AND ITS PREDECESSORS

The novel coronavirus was first found as atypical pneumonia in a cluster of patients from Wuhan city in China at the end of 2019 (Huang et al. 2020; WHO 2020a). By the beginning of 2020, the world knew of its existence. By the end of January, the virus had escaped the Chinese frontier, reaching Taiwan, Japan, Thailand, South Korea, and the U.S., as described in the COVID-19 timeline provided by the World Health Organization (WHO 2020b). The virus spread quickly around the world, turning into the pandemic we are now amidst. Without discriminating economic, geographical, or climatic conditions, up to date, COVID-19 has reached almost all countries around the globe (WHO 2020c).

Nowadays, working with "Big Data" is common in almost every field of knowledge. Daily, every person with a phone, medical record, driving license, or affiliation to an educational institution is providing data to feed multiple databases. However, in such cases, plans have been made to collect and organize the information adequately for storage and further use. On the other hand, in a novel disease outbreak scenario, the time and conditions in which the pathogen will appear and its comportment are unknown, making it impossible to have a ready-to-use system. Yet learning from previous pandemics and contagious disease outbreaks may help us prepare a ground plan if collecting data becomes vital in a short time frame.

The last declared pandemic before COVID-19 was H1N1 influenza, an outbreak first detected in Mexico in February 2009. The H1N1 influenza was caused by a novel influenza A type virus and extended initially to the U.S., Canada, Spain, the U.K., New Zealand, Israel, and Germany before reaching several countries across all continents (Fineberg 2014). Despite its global extension, the estimated deaths worldwide caused by H1N1 influenza in two different studies were 123,000 to 203,000 deaths (Dawood et al. 2012) and 105,700 to 395,600 deaths (Simonsen et al. 2013).

DOI: 10.1201/9781003137993-9

Such data did not differ much from the deaths associated with seasonal influenza in the U.S. However, in contrast to seasonal influenza, H1N1 influenza affected more children and young adults, increasing the disease's severity in terms of years of life lost (Viboud et al. 2010).

H1N1 influenza and its predecessor (the severe acute respiratory syndrome SARS in 2003, which started in China) resulted in more than 8,000 infections and 10% mortality worldwide (LeDuc and Barry 2004)—but they did not reach the infection rate that we are experiencing nowadays. For COVID-19, the current global count is 77,773,667 confirmed cases and 1,710,335 deaths as reported by WHO at 8:52 GMT on December 22, 2020 (WHO 2020c).

Another pandemic worth mentioning is the great influenza pandemic of 1918–1919, considered the deadliest pandemic in the twentieth century. Its origin is unclear, but it spread quickly, infecting about 500 million people worldwide (one-third of the global population) and killing 50 to 100 million people (Johnson and Mueller 2002). One unique feature of this pandemic was the high mortality rate of young people, predominant in those belonging to the 20 to 40 year age group (Centres for Disease Control and Prevention 2019). In that time, the reported data merely documented the situation's extent: there was no approach to use data analysis to understand the infection spread to improve strategies, isolate clusters, and reduce new case numbers.

In 1918, global conditions facilitated the reproduction and spread of the virus—a world in war, soldiers' mobilization, refugee migration, and women engaged in extra-domestic activities—to prepare the perfect conditions to give the influenza a free ticket to travel around the world. But at the time, nobody knew what caused the disease (Martini et al. 2019). However, the contagiousness of the disease became evident, and thus preventive measures became essential. Countries worldwide started implementing public health strategies to reduce the influenza spread (Reid, Taubenberger, and Fanning 2004). After a tremendous death count and three identifiable peaks worldwide, the number of cases started to reduce, and countries began to lift their quarantine measures. The last registered wave was in Japan at the end of 1919: the pandemic had ended by 1920 (Martini et al. 2019).

When the pandemics of 2003 and 2009 appeared, the viral infection's social conditions and characteristics helped stop its spread. The world was not at war anymore, and research on viral infections has progressed dramatically. In contrast to COVID-19, the SARS and H1N1 influenza were less transmissible and with a higher death rate (Petersen et al. 2020), which helped contain the outbreaks. Necessary measures such as quarantines and improved detection in an early contagion phase before patients became vectors proved more effective. In contrast, COVID-19 has displayed a high rate of asymptomatic cases, contagious even without being aware of carrying the disease. This novel scenario represents a challenge for quarantines and containment measures (Zhou et al. 2020; Nikolai et al. 2020).

8.1.2 Data Collection: Past and Now

For the 1919 influenza, data collection was limited to merely reporting the number of infections and deaths. However, for SARS and H1N1, collected data was used to predict future scenarios and even to model the transmission mode (Mutalik 2017; Bauch et al. 2005; Yu et al. 2004). The fast evolution of mobile technologies provides the ability to perform large-scale database analysis, enabling us to speed up data collection in real time. It improves accessibility for scientists and institutions to analyze and publish the current status, predict possible future scenarios, and propose new strategies to prepare the health services and the community.

However, each country faces its own challenges regarding extended experience with data collection and big databases, technology, and resource availability. Developed countries tend to have organized systems to monitor diseases and better preparedness and accessibility to technological resources. On the other hand, developing countries have many challenges to face, since resources destined for technologies and innovation are scarce, and governments lack the human resources and organizational efficiency to design and implement policies to restructure (Cirera and Maloney 2017).

For COVID-19, each country has adopted its measures, ranging from strict quarantines to recommendations to the public, and yet, there is no consensus on which method works better. Adherence to quarantines is difficult, and "smart quarantines" are flawed if the public does not apply the recommendations as much as possible. However, in sectors where poverty afflicts the population, they must balance to stay at home to avoid the virus or going outside to find enough income to feed their families for the day. Proper data analysis, cluster tracking, and clear communication to the public may increase population trust and adherence to the preventive measures. To achieve the mentioned objectives, the government must ensure that the data collection is adequate to use the most updated information. However, when a country is not ready (or does not expect) for such a massive data input in a short time, the data collection and storage get improvised and untidy, and this hinders the data analysis process and delays vital information.

In this chapter, we will introduce the analysis of a real case-scenario on the data collection related to the COVID-19 pandemic in Colombia. One of the authors of this chapter dedicated several months to fix the data provided by the Colombian National Institute of Health (from now on, called INS by its acronym in Spanish). While analyzing the data to give some insights about the country's situation, he realized that the database frequently changed, presenting a challenging scenario for researchers to use such information to report and predict the Colombian status (Manrique Ruiz, Muñetón Santa, and Loaiza Quintero 2020). Next, we will describe the necessary pre-processing required to get an appropriate database for analysis, present some novel data on transition probabilities, and comment on recommendations that could be adopted for future data collection in emergencies.

8.2   DATA AND METHODS

In Colombia, the data regarding COVID-19 is collected, organized, and provided by INS on its website (INS 2020). These files have been anonymized before upload and are open to the general public. The data used for the present analysis includes the files from March 14 to April 30, 2020.

8.2.1   DATA INCONSISTENCIES

Before using the data, its status was confirmed by the authors. Some of the files had several inconsistencies that required important data tiding before creating a database for analysis. These problems are summarized as: (1) lack of verification, (2) poor standardization, and (3) formatting inconsistencies.

8.2.1.1   Data Release Without Verification

The INS started providing daily COVID-19 data from March 14, 2020. An initial inspection of the data suggested that it has been released without verification, since the files contained multiple errors including typos, data origin mislabeling, and patient duplication or merging. The lack of data confirmation led to data loss, accumulated errors, and increased data pre-processing.

8.2.1.2   Poor Standardization of the Collected Data

The initial selection of variables chosen by the INS did not last long. Shortly after the initial files were released, the updated data files' format has been continuously changing. The initial files contained few variables; also, ranges were used for patient identity protection for variables such as age, when the number of cases was not considerable and easy to pinpoint. Later, new variables were added, and the definition of pre-existent variables changed. At some point, the information started to appear as numerical, instead of range, using the same column in the case of age. Other errors, such as date definition and typos also made the data difficult to manage.

8.2.1.3   File Format Change

From March 14 to October 29, 2020, the file format used for daily uploads was Excel spreadsheets (.XLS). However, the large amount of new daily data reached the maximum number of rows in an excel sheet by the end of October. For this reason, the INS decided to change the file format from .XLS into .CSV from this point forward. Since the data was increasing daily, the INS proceeded to compress the information inside RAR containers. Even though the RAR file name was well titled, the compressed information had a simple name (output_open_data.csv) that needs to be taken into account to analyze the information and avoid data loss.

8.2.2   DATA CLEANSING AND PREPARATION FOR ANALYSIS

8.2.2.1   Initial Inspection and Cleansing

The INS' daily spreadsheets contained the anonymized information of COVID-19 positive patients reported in Colombia. The dataset from previous days is updated (as state change and date change), and the new data is added sequentially.

The definition of each variable provided by the INS is explained in Table 8.1. The initial inspection of the data showed some inconsistencies that must be solved before proceeding to the analysis. In an early stage, variables such as age were defined as an interval to protect the patients' privacy. However, it changed to a numeric type at the beginning of April and the variables related to date type. These were initially specified as a date–time, but later the format changed as date only.

TABLE 8.1
Data Definition

Variable	Description	Type
Case ID	Identification	Numeric
Notification date	Date of notification to the National Public Health Surveillance System (SIVIGILA by its acronym in Spanish)	Date
DIVIPOLA code	City code provided by the Political Administrative Division of Colombia	Alphanumeric
City	Colombia's city name	Alphanumeric
Department or district	City's Department	Alphanumeric
Attention	Home, recovered, deceased, Intensive Care Unit (ICU), hospital, Not Applicable (NA)	Alphanumeric
Age	Contains information in intervals and numbers	Alphanumeric
Gender	Male/female	Alphanumeric
Type	Imported, related	Alphanumeric
State	Mild, moderate, severe	Alphanumeric
Country of origin	Patient's country of origin	Alphanumeric
Symptom onset date	Approximate date of starting symptoms	Date
Death date	Date	Date
Diagnosis date	Date of diagnosis by the hospital	Date
Recovery date	Approximated date of recovery	Date
Web report date	Date reported on the website	Date
Recovery type	Polymerase Chain Reaction (PCR), time	Alphanumeric
Department code	Department code of Colombia	Alphanumeric
Country code	Country code where the patient comes from	Numeric
Ethnicity	Native, Afro-American, Raizal, Rom, and Other	Alphanumeric
Ethnic group name	Not available	Alphanumeric
Location recovered	Home, deceased, hospital, ICU, NA	Alphanumeric
Nationality (replaced)	Patient nationality	Alphanumeric

Other problematic variables were the ones containing information from departments and cities. Issues related to typing, accents, and different capitalization patterns, among others, were found. To standardize each department and city's name, we downloaded the information related to the city code from the Political Administrative Division of Colombia (DIVIPOLA by its acronym in Spanish). This information is provided by the National Administrative Department of Statistics (DANE by its acronym in Spanish). Later, we matched the INS records with the DIVIPOLA dataset to find the correct city and department code.

Examining the information further, we found some fundamental issues. For instance, we noticed that a unique case ID could contain data from more than one person when filtered over the time of publication. These errors were identified by comparing cases by publication date, age, gender, and symptom onset date, among other variables.

The data was fixed using a semi-automatic system programmed in a Raspberry Pi 3 server. It was designed to download, standardize, and organize the information while removing noise and correcting the errors mentioned above. Using Python 3.6.8, the INS' daily newsletters were downloaded automatically every day at midnight (Colombia time: UTC -5). After cleansing, the data was ingested into a PostgreSQL database v.12.3.

8.2.2.2 Transitions Correction

In the data released by the INS, there are three states related to location: *home*, *hospital*, and *ICU*. Moreover, two other states are related to condition: *recovered* and *deceased*. The interaction of patients with each location should be logical. The possible transitions between locations considering the health system are shown in Figure 8.1. The states related to the location were considered origins, and the states regarding conditions were considered destinations. A description of severity with different *asymptomatic, mild, moderate,* and *severe* levels was assigned in each location state. The severity levels set to the location *home* are restricted to *asymptomatic* and *mild*, while *hospital* and *ICU* can display *mild, moderate,* and *severe*. Arrows represent the correct directions between location and condition states showing the transit restrictions of the system. For example, a patient in the *ICU* can transition directly to *deceased* but not to *recovered* without passing through the state *hospital*.

After the initial cleansing, the information corresponding to each patient transition was checked, and abnormal behavior was registered for some cases. For instance, patients in the *ICU* went back *home* directly, or patients who appeared *deceased* transitioned to a *recovered* state.

Once the information was fixed and integrated into a database, several queries and dashboards (see Figure 8.2) are made to track the virus' behavior in different cities.

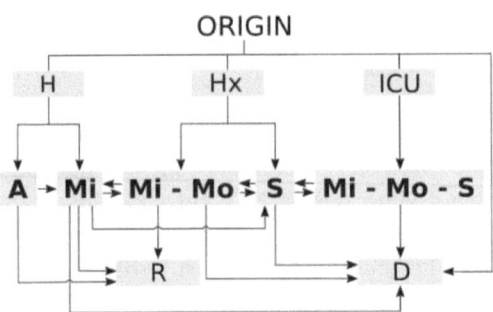

FIGURE 8.1 Visual representation of possible transitions between locations in an origin–destination matrix

Note: H: home; Hx: hospital; R: recovered; D: deceased; A: asymptomatic; Mi: mild; Mo: moderate; and S: severe. Note that the directions of the arrows represent the possible transitions.

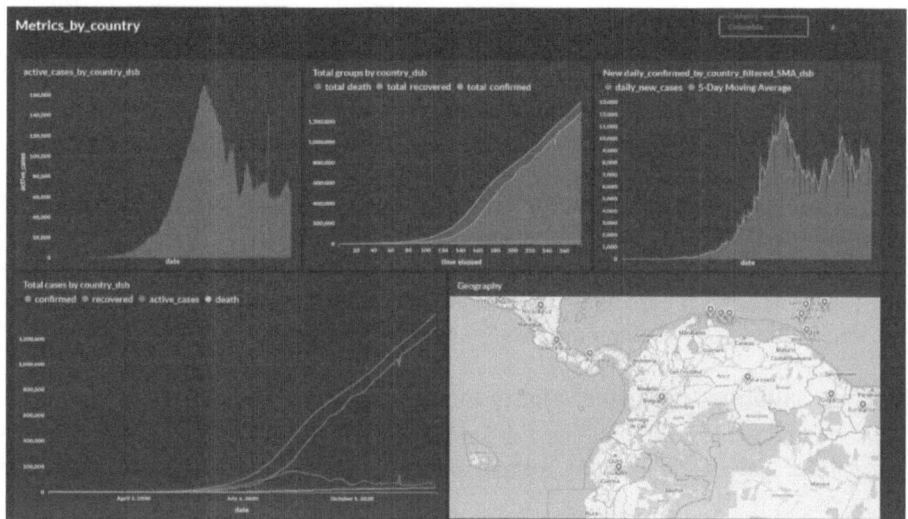

FIGURE 8.2   Dashboard generated with data from the INS, displaying basic epidemiological information

We will now discuss the methods used to correct the information, the use of dashboards, and an investigation on transition probabilities for patients inside the health system (Manrique Ruiz, Muñetón Santa, and Loaiza Quintero 2020).

8.2.3   METHODS FOR DATA CORRECTION

8.2.3.1   K-medoids

The K-medoids or Partitioning Around Medoids (PAM) is a clustering technique that divides the dataset into a *k* number of clusters. It is a more robust version of the K-means (Schubert and Rousseeuw 2019). The objects inside each cluster are similar, and thus the items of other clusters are dissimilar. A cluster's medoid is defined to be where the average dissimilarity to all the cluster objects is minimal (Rousseeuw and Kaufman 1987). If the medoids are found, the data is classified into the nearest medoid cluster (NCSS Procedures 2020). The optimal number of clusters (*k*) can be defined using different techniques such as silhouette analysis (Subbalakshmi et al. 2015).

Pseudo-code (Arora, Deepali, and Varshney 2016)
Input: Let us define the K_i as the number of clusters
     D_i corresponds to the dataset containing *n* objects
Output: A set of k_i clusters.
Algorithm:
1. Randomly select k_i as the medoids for *n* data points.
2. Find the closest medoids by calculating the distance between data points *n* and medoids *k* and mapping data objects.

3. For each medoids *m* and each data point *p* associated with *m,* action the following:
   - Swap *m* and *p* to compute the total cost of the configuration
   - Select the medoids *p* with the lowest cost of the configuration.
4. If there is no change in the assignments, repeat steps 2 and 3 alternatively.

The dissimilarity matrix shows the similarity pair to pair between sets. It is squared and symmetric. The diagonals are defined as zero, where zero is the measure of dissimilarity between an element and itself. The mentioned matrices are calculated with the Gower metric; it calculates the distance between the categorical and numerical attributes (Gower 1971), and the values are between 0 to 1.

8.2.3.2   Silhouette Cluster Validity Index

This method is useful to validate a crisp data cluster (Rousseeuw 1987). It provides a graphical way to analyze how well an object lies inside a cluster. It is calculated as follows:

$$s(i) = \frac{a(i)b(i)}{\max\{a(i),b(i)\}} \qquad \text{2.1}$$

Where *s(i)* represents the silhouette for the object *i*, *a(i)* is the average dissimilarity of the ith data point, *b(i)* is the minimum average dissimilarity of the ith data point to another cluster where *i* is not a member. *S(i)* is then between -1 and 1.

8.2.3.3   Transition Matrix

As described by Manrique Ruiz, Muñetón Santa, and Loaiza Quintero (2020), the stochastic matrixes are squared matrixes that detail changes between states. These transitions are calculated as the probabilities of transit between state *i* to state *j*. In the case of a previous investigation, we estimated this matrix daily.

$$P(j|i) = P(i,j) \qquad \text{2.2}$$

The sum of all probabilities for the state *i* must be 1.

$$\sum_{i-1}^{m} P(i,j) = 1 \qquad \text{2.3}$$

The states for this matrix are then *home* (H), *hospital* (Hx), *UCI, recovered* (R), and *deceased* (D). It is easy to identify *deceased* as an absorbent state, defined as:

$$P(x_{t+1} = i, x = i) = 1 \qquad \text{2.4}$$

Where once a patient reaches this type of state, he cannot leave it.

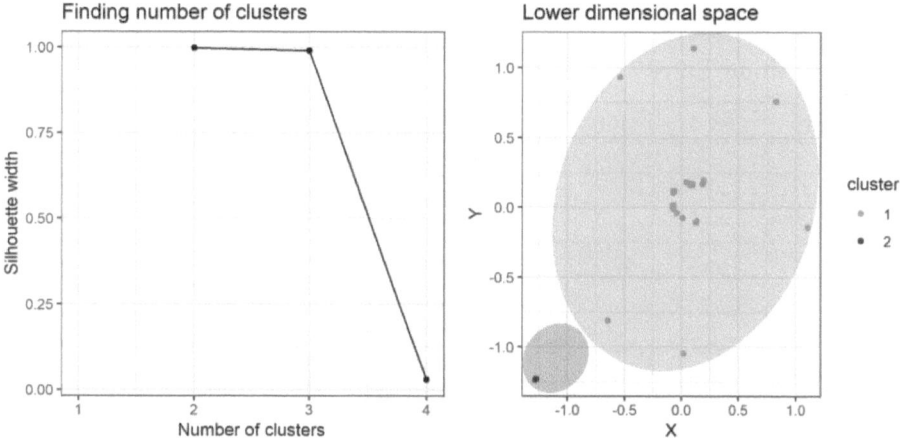

FIGURE 8.3 Finding the number of clusters: *(left)* showing the optimal number of clusters by using the Silhouette width; and *(right)* showing the visualization of these clusters by using a lower-dimensional space

8.3 RESULTS

The clusters of unique records (*new_case_id*) were estimated using the DIVIPOLA's city code, gender, age, and diagnosis date. The processing and analysis were conducted in Microsoft R-open v.3.5.1 (64-bit).

Figure 8.3 shows an example from a single patient record when the optimal number of clusters was found through the silhouette width.

The variables' weights were [0.1, 0.4, 0.4, and 0.1] for the DIVIPOLA code, gender, age, and diagnosis date.

8.3.1 CONFIRMATION OF TRANSITIONS THROUGH DYNAMIC WINDOWS

Once the *new_case_id* is identified through clustering, we check the states' information where a person with coronavirus could go. By filtering per new_case_id, different graphs were found, for example:

1. "*home – hospital – hospital (ICU) – deceased*"
2. "*home – recovered – deceased – recovered*"
3. "*hospital – hospital (ICU) – home*"

In the Case 1, the found pattern agrees with the health system transition logic. However, there are logic issues for Cases 2 and 3. In Case 2, a person in *recovered* state transited to *deceased* and back to *recovered*. Logic indicates that it was a data input mistake, where *deceased* should have never existed. For Case 3, the sense implied in the health system was overlooked. A patient can enter directly into the *ICU*; however, it is impossible to transit straight to home from *ICU* once the patient's health improves. A down step to *hospital* is required before discharge, and so this was also considered a mistake.

TABLE 8.2
Fixing Transition States

State (Raw)	State (Fixed)
home	home
home	home
home	home
hospital	home
home	home
hospital	hospital
hospital	hospital
hospital	hospital
hospital (ICU)	hospital
hospital (ICU)	hospital (ICU)
hospital	hospital (ICU)
hospital (ICU)	hospital (ICU)
hospital (ICU)	hospital (ICU)
deceased	hospital (ICU)
deceased	deceased
home	deceased
deceased	deceased
deceased	deceased
deceased	deceased
deceased	deceased

Note: Example of a patient over 20 days of evolution. The fixed state is calculated using a median of t = -3days.

In the Case 1, we calculated the median in dynamic windows of time $t = -3$ days. By using this method, we can fix the writing issues. Table 8.2 shows an example of transitions correction. The left column shows the raw data, while the right one shows the fixed transitions through dynamic windows.

8.3.2 TRANSITION PROBABILITIES

The authors wanted to analyze Colombia's elderly population affected by COVID-19, who are considered the most susceptible to this disease. In the document published by Manrique Ruiz, Muñetón Santa, and Loaiza Quintero (2020), three age ranges were defined as follows: 0–24 years old, 25–65 years old, and 66 years old and more. Figure 8.4 shows the graph of transition probabilities for patients over 66 years old in Bogotá.

The dataset used in this chapter does not contain reports of re-infections. For that reason, the state *recovered* cannot be an absorbent state. The reader can see in Figure 8.4 that a person older than 66 years old in Bogotá can transit from *recovered* to *home* with a probability of 40%. However, some patients appeared as *recovered* to

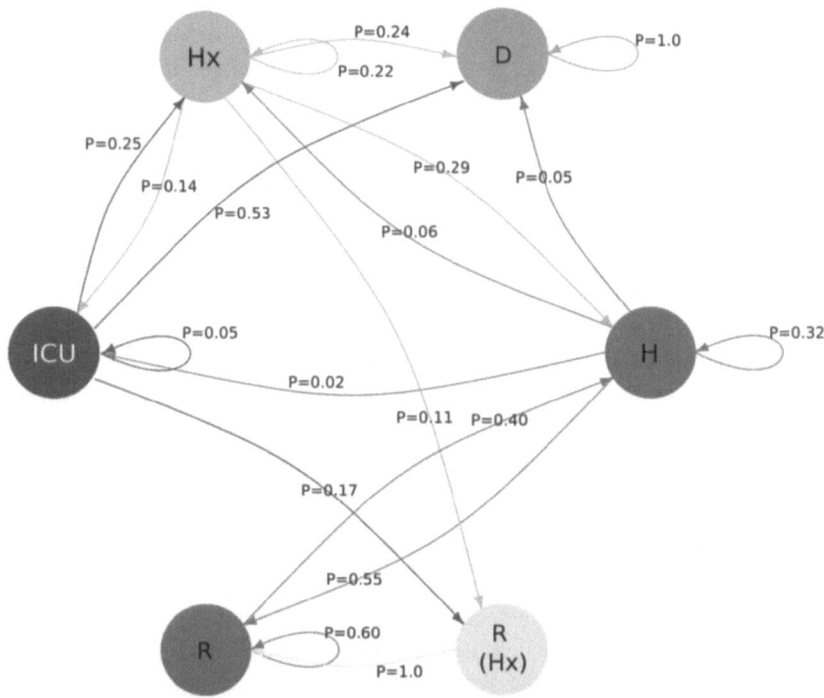

FIGURE 8.4 Example of transition probabilities for patients in Bogotá who are older than 66 years

Note: H: home; Hx: hospital; R: recovered; and D: deceased. The term *P* represents the probability of going from state *i* to *j*.

hospital since after recovering, complications from the COVID-19 infection or pre-existent conditions required treatment before the patient could be discharged home.

8.4 DISCUSSION

When this chapter was written, several investigations on COVID-19 were ongoing in multiple fields such as epidemiology, immunization, and virus transmission, among others. Up to date, almost all countries are releasing their COVID-19 information as open sources to facilitate scientists to provide fast and updated solutions for the progressing pandemic. Several scientists and research institutions have been using such data to answer the novel and constantly changing scenario that the pandemic has brought worldwide in 2020.

The availability of complete and reliable data is essential to offer quick answers to reduce human losses, prioritize resources, adopt preventive measures, and update in-use policies due to the situation's continuous changes. This chapter's specific scenario aimed to highlight potential and common mistakes during data collection, and to reflect the increased time that pre-processing takes when data is not collected with high standards.

8.4.1 STRATEGIES FOR IMPROVING DATA COLLECTION

The pandemic outbreak has brought several unique challenges for control and management of the situation; data analysis has become a useful tool in the fight to curb new cases. Analyzing the data released by the Colombian government, we noticed the lack of consistency across several variables and some errors that led to increased time spent in data cleansing rather than in data analysis. Next, we will provide some insights according to our experience.

8.4.1.1   Variable Definition

To unify and make the data easy to use and accessible to researchers, it is crucial to consider what kind of variables must be stored and, perhaps in the future, what new variables may be integrated. Scientists such as Xu and colleagues (2020) propose to record a large set of variables containing demographic, geographic, disease-related, and identification-related information. Their variable selection seems complete; however, they propose that some variables may contain multiple formats depending on the information's availability (Xu et al. 2020). We disagree with this approach. In our own experience, we found multiple formatting cases for a single variable in the data released by the INS in Colombia. As mentioned before, during the early stage of data collection in Colombia, the "age" variable was defined as ranges to protect the patients' identity. However, the definition of this variable changed to numeric after the number of patients became larger. We consider that the definition of a variable should be the same through all the data collection process. The integration of multiple formats for a single variable should not be implemented since the data preparation for future investigations will become time consuming and will be more likely to contain errors. For that reason, we propose that the original "age" variable defined as range should be preserved and a new variable should be added but containing only numerical data. The name of the new variable must differ from the original one. Even though it is crucial to anonymize personal information, some techniques could be useful in different cases, such as hashing variables. The hash functions were introduced in cryptography to protect information's authenticity (Mohanty, Sarangi, and Bishi 2010).

In date–time variables, several formats, including date and time, only date, and date in different patterns, are available. However, the variable must be defined using only one format. Formats such as YYYY-MM-DD(TZ) are well known and easy to understand. Suppose time must be included because it adds important information. In that case, it could be included in the same variable, or a new one can be created with the time information only: HH:MM:SS.000(TZ). Additionally, the time zone (TZ) could be essential to unify the collected data easily with information from other sources. Nevertheless, the time zone can also be derived from the source of information.

8.4.1.2   File Naming for Storage

When automated systems are developed, unified file names are required to access the information correctly and quickly. In the pandemic scenario where data files are released daily, the names of the files should be consistent and easy to identify. The

Open Data Institute reported acceptable data practices (ODI 2020), giving an example of proper file naming as follows:

- http://mysite.org/data/2020-04-19.csv
- http://mysite.org/data/2020-04-20.csv

In this scenario, files are easy to locate by the published date. Sequential numbering or sequences of letters should be avoided since names such as "day 240" are not easy to locate in a timeline. Since November 2020, the INS changed the type of file, and the names changed from Excel spreadsheets (XLS) to comma-separated values (CSV).

8.4.1.3   File Type and Properties

There are several file formats, but one of the most commonly used is CSV. Since it is a text file separated by commas, it does not have row limitations. As we discussed before, Excel files have a limit, and thus some information could be lost. The Colombian case is not an isolated one since it is known that other countries, such as the U.K., lost several thousands of COVID-19 data records due to this constraint (Linder 2020; Landler and Mueller 2020).

It is also essential to declare the encoding to read all the content without losing or having problems with special characters. One of the most common encodings that could be appropriate for some countries is UTF-8. In cases of bulk data collection, the data files may become very large and difficult to store. A good alternative is to provide the information in containers. Some of the most commonly used formats are TAR and GZIP (Stanford Library, n.d.).

8.4.1.4   Missing Data

When dealing with database management, it is vital to consider the presence of null values. It is essential to define them clearly to avoid problems at a later date. It is suggested to standardize all the coded and null values (Axiom Data Science 2017) and use explicit values for missing or no data. Some databases can properly manage missing data defined as NULL or NaN (Not a Number). Neglecting or failing to define properly missing values will lead to errors in further analysis.

8.4.2   Data Cleansing Techniques

The data provided by the INS required revision, cleansing, and standardization before using it in a database. The methods used to correct the information were appropriated since it was possible to integrate different data types. Other heuristics could have been applied; however, the clustering analysis through K-medoids effectively found similarities among the records.

Following Arora, Deepali, and Varshney (2016), the K-medoids worked better than K-means when it is applied to Big Data. Other benefits of using this method are reduced execution time, non-sensitivity to outliers, and noise reduction.

8.5   FINAL NOTE

When collecting data from an epidemic outbreak, it is required for the governments to release the most appropriate information to help future investigations. Also, contingency plans are needed to face fast-moving emergency situations. We hope this chapter may help to build protocols for data collection in emergencies without considering the type of country where it will be applied. Proper data collection speeds data analysis and helps scientists to provide quick answers using the most up-to-date and reliable data.

REFERENCES

Arora, P., Dr. Deepali, and S. Varshney. 2016. "Analysis of k-Means and k-Medoids algorithm for Big Data." *Procedia Computer Science,* 78: 507–512.

Axiom Data Science. 2017. "Data Management Best Practices." www.axiomdatascience.com/best-practices/DataManagementBestPractices.html.

Bauch, C. T., J. O. Lloyd-Smith, M. P. Coffee, et al. 2005. "Dynamically modeling SARS and other newly emerging respiratory illnesses: Past, present, and future." *Epidemiology,* 16 (6): 791–801.

Centres for Disease Control and Prevention. 2019. "History of 1918 Flu Pandemic | Pandemic Influenza (Flu) | CDC." www.cdc.gov/flu/pandemic-resources/1918-commemoration/1918-pandemic-history.htm

Cirera, X. and W. F. Maloney. 2017. *The Innovation Paradox: Developing-country capabilities and the unrealized promise of technological catch-up.* Washington DC: World Bank Publications.

Dawood, F., A. D. Iuliano, C. Reed, et al. 2012. "Estimated global mortality associated with the first 12 months of 2009 pandemic influenza A H1N1 virus circulation: A modelling study." *The Lancet Infectious Diseases,* 12 (9): 687–695.

Fineberg, H. V. 2014. "Pandemic preparedness and response — Lessons from the H1N1 influenza of 2009." *New England Journal of Medicine,* 370 (14): 1335–1342.

Gower, J. C. 1971. "A general coefficient of similarity and some of its properties." *Biometrics,* 27 (4): 857–871.

Huang, C., Y. Wang, X. Li, et al. 2020. "Clinical features of patients infected with 2019 novel coronavirus in Wuhan, China." *The Lancet,* 395 (10223): 497–506.

INS. 2020. "Datos Abiertos Colombia." La plataforma de Datos Abiertos del Gobierno Colombiano. www.datos.gov.co/ (accessed December 16, 2020).

Johnson, N. P. and J. Mueller. 2002. "Updating the accounts: Global mortality of the 1918–1920 'Spanish' influenza pandemic." *Bulletin of the History of Medicine,* 76 (1): 105–115.

Landler, M. and B. Mueller. 2020. "In U.K.'s test and trace: Now you see 'em, now you don't," in *The New York Times* online edition. www.nytimes.com/2020/10/05/world/europe/uk-testing-johnson-hancock.html (accessed December 1, 2020).

LeDuc, J. W. and M. A. Barry. 2004. "SARS, the first pandemic of the 21st century." *Emerging Infectious Diseases,* 10 (11): e26.

Linder, C. 2020. "The U.K. lost 16,000 COVID cases because it doesn't understand Microsoft Excel," in *Popular Mechanics.* www.popularmechanics.com/technology/a34274176/uk-coronavirus-excel-spreadsheet-lost-cases/ (accessed October 6, 2020).

Manrique Ruiz, L. C., G. Muñetón Santa, and O. L. Loaiza Quintero. 2020. "Transiciones entre los estados de diagnóstico de personas con la COVID-19 en Colombia." *Revista Panamericana de Salud Pública,* 44 (171).

Martini, M., V. Gazzaniga, N. L. Bragazzi, et al. 2019. "The Spanish influenza pandemic: A lesson from history 100 years after 1918." *Journal of Preventive Medicine and Hygiene,* 60 (1): E64–67.

Mohanty, R., N. Sarangi, and S. Bishi. 2010. "A Secured Cryptographic Hashing Algorithm." https://arxiv.org/ftp/arxiv/papers/1003/1003.5787.pdf

Mutalik, A. V. 2017. "Models to predict H1N1 outbreaks: A literature review." *International Journal of Community Medicine And Public Health,* 4 (9): 3068–3075.

NCSS Procedures. 2020. "Chapter 447: Medoid partitioning." https://ncss-wpengine.netdna-ssl.com/wp-content/themes/ncss/pdf/Procedures/NCSS/Medoid_Partitioning.pdf.

Nikolai, L. A., C. G. Meyer, P. G. Kremsner, et al. 2020. "Asymptomatic SARS coronavirus 2 infection: Invisible yet invincible." *International Journal of Infectious Diseases,* 100 (November): 112–116.

ODI. 2020. "Publishing Open Data in Times of Crisis." Open Data Institute. https://drive.google.com/file/d/1rJDRk8KSrMsx7kB4B6_tPfaau6JslCKQ/view.

Petersen, E., M. Koopmans, U. Go, et al. 2020. "Comparing SARS-CoV-2 with SARS-CoV and influenza pandemics." *The Lancet Infectious Diseases,* 20 (9): e238–244.

Reid, A. H., J. K. Taubenberger, and T. G. Fanning. 2004. "Evidence of an absence: The genetic origins of the 1918 pandemic influenza virus." *Nature Reviews Microbiology,* 2 (11): 909–914.

Rousseeuw, L. and P. J. Kaufman. 1987. "Clustering by Means of Medoids." www.researchgate.net/profile/Peter_Rousseeuw/publication/243777819_Clustering_by_Means_of_Medoids/links/00b7d531493fad342c000000.pdf.

Rousseeuw, P. J. 1987. "Silhouettes: A graphical aid to the interpretation and validation of cluster analysis." *Journal of Computational and Applied Mathematics,* 20: 53–65.

Schubert, E. and P. J. Rousseeuw. 2019. "Faster K-medoids clustering: Improving the PAM, CLARA, and CLARANS algorithms." *Lecture Notes in Computer Science,* 171–187.

Simonsen, L., P. Spreeuwenberg, R. Lustig, et al. 2013. "Global mortality estimates for the 2009 influenza pandemic from the GLaMOR Project: A modeling study." *PLOS Medicine,* 10 (11): e1001558.

Stanford Library. n.d. "Best Practices for File Formats." https://library.stanford.edu/research/data-management-services/data-best-practices/best-practices-file-formats.

Subbalakshmi, C., G. R. Krishna, S. K. Rao, et al. 2015. "A method to find optimum number of clusters based on fuzzy silhouette on dynamic data set." *Procedia Computer Science,* 46: 346–353.

Viboud, C., M. Miller, D. R. Olson, et al. 2010. "Preliminary estimates of mortality and years of life lost associated with the 2009 A/H1N1 pandemic in the US and comparison with past influenza seasons." *PLoS Currents,* 2 (March).

WHO. 2020a. "Pneumonia of Unknown Cause – China." Emergencies Preparedness, Response. www.who.int/csr/don/05-january-2020-pneumonia-of-unkown-cause-china/en/.

WHO. 2020b. "Listings of WHO's Response to COVID-19." www.who.int/news/item/29-06-2020-covidtimeline (accessed November 27, 2020).

WHO. 2020c. "WHO Coronavirus Disease (COVID-19) Dashboard." https://covid19.who.int (accessed November 27, 2020).

Xu, B., B. Gutierrez, S. Mekaru, et al. 2020. "Epidemiological data from the COVID-19 outbreak, real-time case information." *Scientific Data,* 7 (1): 106.

Yu, I., Y. Li, T. W. Wong, et al. 2004. "Evidence of airborne transmission of the severe acute respiratory syndrome virus." *New England Journal of Medicine,* 350 (17): 1731–1739.

Zhou, F., J. Li, M. Lu, et al. 2020. "Tracing asymptomatic SARS-CoV-2 carriers among 3674 hospital staff: A cross-sectional survey." *EClinicalMedicine,* 26 (September): 100510.

9 An Agent-based Methodology for Seaport Decision Making

Ana X. Halabi-Echeverry, Nelson Obregón-Neira,
Hugo L. Niño-Vergara, Juan C. Aldana-Bernal,
and Milton Baron-Perico

9.1 INTRODUCTION

Fundamentally, agents are automata inspired by humans. The construction of an agent-based system becomes part of the foundations of Artificial Intelligence (AI) (Russell and Norvig 1995). The interaction among agents is understood as a multi-agent behavioral system. Benenson and Torrens (2004) define a multi-agent system (MAS) as bounded agents situated in an environment. In this chapter, environments shall be understood as the universe of discourse R able to represent the knowledge of the domain so that an automated reasoner can represent problems and generate solutions to these problems. To develop such a system, one necessary course of action is to allow an agent to know about their circumvent environment. Intuitively the importance of knowledge and experience in making good decisions is obvious. We focus our research on the integration of *knowledge management* representations and *simulation techniques* to address our goal to support seaport decision making. Technical properties and challenges in the seaport domain set up a continuous research effort on more intelligent support to guide expertise (Murty et al. 2005). Technical properties can be considered as the new capabilities of seaport collaboration in which development of capabilities in sharing information, planning, and execution allow two or more seaports to advance.

This chapter presents an agent-oriented methodology in this domain to answer the following questions:

- What are possible considerations to develop an agent-based methodology invoking a collaborative environment for seaport authorities?
- What is the knowledge/epistemological level that may describe what the simulated agents need to know?

DOI: 10.1201/9781003137993-10

9.2 COMPLEXITY OF THE DECISION-MAKING ENVIRONMENT IN SEAPORTS

In agreement with Bennet and Bennet (2008), a definition of complexity in a decision-making general context is given as:

> Complexity is the condition of a system, situation, or organisation that is integrated with some degree of order, but has too many elements and relationships to understand in simple analytic or logical ways ... In the extreme, the landscape of a complex situation [system] is one with multiple and diverse connections with dynamic and interdependent relationships, events and processes.
>
> *2008, p. 5*

According to Bennet and Bennet, the decision/problem space R for complex systems should include (2008, p.5):

- Perceived boundaries of the system.
- The ontology of the situation.
- Sets of relative data and information.
- Observable events, history, trends, and patterns of behavior.
- The underlying structure and dynamic characteristics.
- The identity of the individuals/groups involved.

By *perceived boundaries of the system,* it is meant the forces controlling/influencing the boundaries of the system. Complex systems are sensitive to those forces because they impact the system's behavior. The *ontology of the situation* represents a set of characteristics surrounding the decision strategy that have an important influence on the desired outcome. Regarding the *underlying structure and dynamic characteristics* of a complex system, it is desirable to understand how the system operates and what it takes to resolve problems. Considering seaports as complex systems and as socio-technical systems allows a framework to consider phenomena such as self-organisation and the emergence of highly changing, non-linear dynamics, and heterogeneous or anisotropic processes out of balance.

In the same line, decisions are complex outputs of a managerial process that differ from one problem to another. Land-use constraints, international normativity and pressures, and local/global economic crisis are some examples of complex situations in which seaport decision makers operate and undertake decisions daily. Pomerol and Adam (2006) state that "intelligence" complements the part of the decision that is unstructured, infrequent, or novel. While Bennet and Bennet ascertain that non-programmable decisions dealt using a problem-solving approach may,

> [F]ind the cause of the problem, change it, and the problem go away [but that] does not work if the problem is complex, [because] what typically happens is that the change works for a short time and the complex system rejects it and the problem returns, larger than before.
>
> *2008, p. 12*

Increasingly, decision-making effectiveness relies on how decisions are made. Stabell, cited in Gachet and Haettenschwiler (2006), proposes a classical decision-oriented approach for decision support development that distinguishes between "substance" (what is used during decision making) and "procedure" (how decisions are made), which can also be extended to other questions such as "how decisions should be made" or "not made as they should be". Considering why and how decisions are made guides to the structure and dynamic characteristics of the complex system. Stabell's approach involves three interrelated activities to ensure a correct system's development (Gachet and Haettenschwiler 2006, p. 103):

1. Data collection: Including data on current decision making using various techniques (interviews, observations, questionnaires, and historical records).
2. Descriptive modeling: Establishing a coherent description of the current decision process.
3. Normative modeling: Specifying a norm for how decisions should be made.

The data collection is an important challenge in current computational developments because it may increase the intelligence of knowledge-based systems (KB) through an appropriate data gathering, storage, and knowledge management that enables an input to the decision process (Negash and Gray 2008). Descriptive modeling emphasis is upon learning from observation of real-world decision making, and normative modeling upon analytical tools, model building, and the scientific method. The challenge posed by Bennet and Bennet determines that in the approach to complex systems,

> [It] remains to be seen how or if it is possible to take complex situations and identify these separate aspects (i.e., boundaries of the system; ontology of the situation; relative data and information; observable events, history, trends, and patterns of behavior; underlying structure and dynamic characteristics; and individuals/groups involved) of the system in such a way that one could choose the most effective decision strategy.
>
> *2008, p.18–19*

To allow a comparison with Bennet and Bennet's approach, DeTombe (2001, 2002, 2013a, 2013b) presents additional elements of analysis about the individuals/groups involved in complex decision environments. These elements are; knowledge, power, and emotion. *Knowledge* is understood as a way of building solutions to problems using expertise and disciplinary understanding. *Power* deals with the way decision makers can stimulate, delay, or prevent solutions. *Emotion* is understood as human aspects of problem handling, such as hidden agendas and intuition. DeTombe (2001, 2013b) also refers to the human level of effort to change a situation (namely, micro-level for minor changes, meso-level for major changes, and macro-level for exceeded efforts) and to the dimensions of the decision consistent to the technical–human element of analysis. Finally, Detombe's framework considers temporary situations and ways in which the complex decision is affected. The following five elements summarize DeTombe's general framework of complex decision making:

1. Levels at which to handle the decision.
2. Dimensions of the decision.
3. Experts and stakeholders involved in the decision.
4. Temporary situations of the decision.
5. Ways in which the decision is connected: cause and effect influences.

9.3   THE NEED FOR A METHODOLOGY TO SUPPORT SEAPORT DECISION MAKING

In a complementary direction, Cassaigne and Lorimier (2006) observe the complexity of the decision supported, i.e., decision maker and the expert knowledge (which sometimes does not reside in the decision maker) and the intelligent support. Figure 9.1 provides the main elements and roles of interaction among the parts

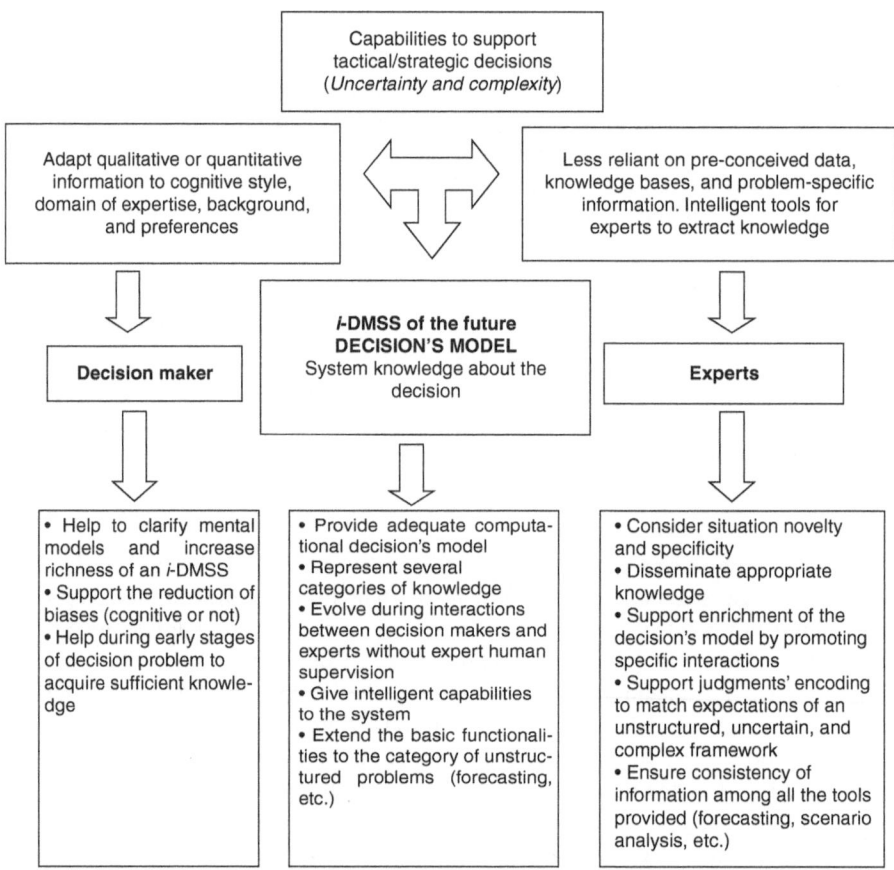

FIGURE 9.1   Elements and main roles of interaction among the parts involved in an intelligent decision-making process

Source: Adapted from Cassaigne and Lorimier, 2006, p.402.

involved in a decision-making process. As seen in the middle section, a decision's model shall represent several categories of knowledge and "it is a major challenge for future *i*-DMSS [intelligent support] to propose an adequate framework, sufficiently flexible to evolve without requesting expert human supervision" (p.417).

To find sustainable solutions to the complex challenges, seaport authorities aim at regulating the decisions within the port [ecosystem], specifically including controlling, surveillance, and policing functions in view of ensuring safety and security within the port, but also concerning environmental protection (Acciaro et al. 2014). From this, knowledge and information sharing can be understood as a way of building solutions to problems using that expertise and disciplinary understanding of collaborative decision making. Holsapple, cited in Chi et al. (2008), defines *collaboration* as,

[A]n interactive, constructive, and knowledge-based process, involving multiple autonomous and voluntary participants employing complementary skills and assets, with a collective objective of achieving an outcome beyond what the participants' capacity and willingness would allow them to accomplish individually.

2008, p.452

Chi and colleagues supplements the previous definition with the following axioms (2008, p. 452):

- Collaboration is episodic, involving episodes of varying durations that may be linked in varying patterns.
- Collaboration requires an internal governance structure, which can range from the rudimentary to the complex and can have formal and informal aspects.
- The internal governance structure of a collaboration episode includes both infrastructure and culture.
- The process and outcome of a collaboration episode are influenced by the environment within which it occurs.

Concepts on collaboration, inter-organisational systems (IOs), and network organisations are all used in collaborative decision making in seaports. Seaport authorities as decision makers must base their standpoints on different scenarios (e.g., through simulations) suitable to show outputs against empirical decisions (as shown in Figure 9.1). The current available simulation framework in literature shows at least three approaches for seaport decision making: (1) agent-based models, (2) system dynamics models, and (3) optimisation and statistical models.

A brief literature review of these methods shows that agent-based models are the most common models for planning the use of resources, such as land. Han et al. (2009) states that agent-based simulation allows modeling of both (static) structure and (dynamic) behavior. Therefore, because of the inherently dynamic and reactive behavior of agents, it is possible to capture and reveal temporal changes that occur in the landscape. Other authors, such as Lee, Song, and Ducruet (2008), consider agents or multi-agents modeling as a theoretical framework. On the other hand, Wiek and Walter (2009) define a structured multi-agent model as combining system knowledge and negotiation to cope with cross-sectoral planning.

From a more systemic perspective, authors such as Luan, Chen, and Wang (2010) investigate the impacts of various port-related driving forces on a city. System dynamics models are well known to provide a comprehensive approach to model complex systems. Therefore, Luan, Chen, and Wang (2010) used system dynamics models to develop a model of the port–city system within an international economic context. Another approach, followed by Georgakaki et al. (2005) and Giannouli et al. (2006), promotes the use of simulating dynamic behavior when monitoring the traffic activity, and has developed a series of modules to enable the collection and compara-tive assessment of traffic emissions.

Finally, optimisation and statistical models are potentially suitable techniques to apply to port data, and are particularly relevant for research that focuses on scheduling, routing vessels, optimisation of shipping routes, and similar issues. Sirikijpanichkul et al. (2007) studied the minimal social costs or externalities when assessing a new freight hub location. Their modeling process is based on inputs, outputs, and constraints of individual decision makers, which in turn constitutes one hybrid model with an agent-based approach. Others, such as Tongzon, Chang, and Lee (2009) and Yeo, Roe, and Dinwoodie (2008), give importance to underlying factors capable of providing empirical evidence for making sound decisions regarding management of ports. Moreover, these authors pay attention to assessing competitiveness and effi-cient performance of ports, by relying on the statistical modeling techniques.

9.4   IS AGENT-BASED METHODOLOGY THE KEY?

Agent-based modeling (ABM) has been used as a framework to simulate complex adaptive systems (CAS) for its emergent behavior and a wide range use of domains (Cardinot et al. 2019). Hellman, cited in Russell, Norvig, and Davis (2010), notes that AI is the automation of activities associated with human tasks such as deci-sion making. The meaning of automation also involves aspects of inter-operability. According to Noff (2009):

> Increasingly; this ability [inter-operability] requires cooperation (sharing of information and resources) and collaboration (sharing in the execution and responses) with other devices and systems.
>
> *2009, p.39*

Thus, inter-operability addresses the need for cooperation, exchanging information, and sharing and re-using information. The systems inter-operability is a challenge posed for inter-organisational systems in seaports. New technologies are meant to enable information exchange, planning at a higher level after the exchange of infor-mation, real-time chains, and seamless communication between stakeholders. The major problem arises from the complexity of modeling such an inter-operability process.

Benali and Ben (2011) states that:

> [S]imulation enables engineers to understand the complexity of the system being developed and at the same time to examine how the strategic decisions would influence the overall performance of this system.
>
> *2011, p.133*

From a practical modeling standpoint, work must be done on showing the component's behavior of a system to be able to learn from prior knowledge for better decision making in the port domain, and that can be based on different scenarios (i.e., through simulations).

To observe this kind of system may require, for example, the use of agent-based simulations that would allow decomposing the system into smaller and simpler parts, which are easier to understand, model, and validate. For instance, Singh (2011) declares the use of interaction-oriented programming (IOP), concerned with the engineering of systems comprising two or more autonomous and heterogeneous components or agents, commonly used in applications such as cross-organisational business processes. The agent-based modeling allows the generation of a simulated emergent behavior, which could be observed in the dynamics of the seaport given its consideration as a complex socio-technical system.

9.5   SPECIFYING AN INTERACTION/COMMUNICATION PROTOCOL IN AN AGENT-BASED MODEL

Wooldridge, Jennings, and Kinny (2000) propose the characteristics that agents should have: social ability, autonomy, reactivity, and proactivity. According to Ferrari, Montanari, and Tuosto (2003), the agent paradigm essential properties are:

- An agent, in order to run, needs an execution environment.
- An agent is autonomous: it executes independently of the user who created it (goal driven).
- An agent can detect changes in its operational environment and can act accordingly (reactivity and adaptivity).

9.5.1   PROPERTIES OF AN AGENT-BASED SEAPORT DECISION MAKER

To elaborate, key properties to investigate are agents' active or passive behavior. If active, they tend to initiate their actions to achieve their internal goals; otherwise, when passive, they end to react to other agents and the environment (Macal and North, 2010). Prior work of Halabi-Echeverry (2016) states that seaports may be classified as initiative partners because their international influence and domestic trade volumes are the primary beneficiaries of interaction mechanisms. While a second group of seaports, comprised mainly of small- and medium-sized ports, may be classified as passive partners since mechanisms for seeking collaboration with neighboring ports are unnecessary, increasing their competitor's chances of survival in an already competitive environment. In this chapter, we will use this classification to further explain the interaction and communication protocols necessary to describe an agent-based methodology in this domain.

Possible behaviors of an agent-based seaport decision maker are: (1) their autonomous capability, and (2) their jurisdictional heterogeneity. As mentioned, the task of collaboration with other parties will depend on the governance and management applied in each port and the networking capabilities of their decision makers.

Strategies of each market player may create an uncertain environment in which a weak port authority position not only risks the autonomy of the port, but any independent policy pursued to sustain the port. Secondly, the jurisdictional heterogeneity influences port collaboration mechanisms and stimuli due to distinct economic structures and other singularities in each port.

9.5.2 MULTI-AGENT INTERACTION AND COMMUNICATION PROTOCOLS

In multi-agent interactions, the power of software agents comes from inter-operability, which is achieved through an agent communication language or protocol. Each protocol has a scope within which is the agent role, parameters, and messages, to accomplish its characteristics of: good definition, unambiguous, and formal representation (semantic language).

Interacting agents where each agent has different knowledge about the environment are described by logics about knowledge. Modal logics are often used in this context, since they describe in natural language different modes of truth (necessarily true, true in the future, and so on), which contrasts with classical logic.

According to Chellas (1980):

> The idea is that different things may be true at different possible worlds, but whatever holds true at every possible world is necessary, while that which holds at least one possible world is possible.
>
> *1980, p.3*

We describe two possible frameworks for multi-agent interaction and communication protocols (at the knowledge level to support the dynamic behavior needed for the interactions): (1) IEEE-FIPA (Foundation for Intelligent Physical Agents), and (2) BSPL (Blindingly Simple Protocol Language).

9.5.2.1 IEEE-FIPA

At the international level, there is the IEEE-FIPA, which establishes a common and generic framework for MAS. The first purpose of this architecture is to foster inter-operability and re-usability of the agent specification. The FIPA establishes a guide about the elements that MAS must integrate. Through a semantic and meaningful message exchange between agents, different messaging transports, different agent communication languages, or different content languages can be inter-operable (FIPA 2002). Each of the FIPA directives is defined by:

- The summary where the meaning of the message is briefly explained.
- The content of the message detailing what type of content should be included.
- The description that is a detailed explanation of the communicative act.
- The formal model which is a description in SL (Semantic Language).
- An example of a message with the communicative act.

The FIPA specifications and standards are: (1) Device Ontology Specification, (2) FIPA Nomadic Application Support Specification, and (3) FIPA SL Content

Language Specification. The corresponding FIPA standard is found in SC00008, SI00014, and SI00091.

9.5.2.2 BSPL

Singh (2011) explains that traditional approaches for specifying a simple protocol can be of two orders—declarative or procedural:

1 *Declarative* approaches for meaning:
   • improve flexibility, but
   • under-specify enactment (with potential of inter-operability failures).
2 *Procedural* or declarative approaches for operations are:
   • operationally clear, but
   • tend to emphasize control flow
   • tend to over-specify operational constraints
   • yield non-trivial inter-operability and endpoint projections.

Singh claims that his approach only requires two main constructs: (1) defining a message schema, and (2) composing existing protocols. Accordingly, the BSPL offers:

> [N]o constraints on the ordering or occurrence of messages, deriving any such constraints from the information specifications of message schemas ... [and] captures a variety of common and subtle protocols.
>
> *2011, p.1*

We acknowledge that the BSPL principles are close to the methodological needs defined in our approach, which uses two main types of data. The first is historical, factual, or declarative knowledge (i.e., what data is used during decision making: for example, port performance indicators and regulatory frameworks). When the knowledge of a domain is represented in a declarative formalism, the set of objects that can be represented is called the universe of discourse (Gruber 1995). The second is procedural knowledge (i.e., data about how decisions are made). Both types of data are needed for decisions concerning seaport decision making.

9.5.3 THE KNOWLEDGE/EPISTEMOLOGICAL LEVEL OF AN AGENT-BASED BEHAVIOR

In a search for clear examples that can provide knowledge on types of data and advantages of implementing the agent-based methodology, we will now discuss a real Port State Control (PSC) context.

The Operational Network of Regional Cooperation among Port Authorities of Latin America is a proper mechanism for data sharing and collaboration between maritime and port authorities. Using some of their empirical evidence, we indicate agents' needs for increasing the compliance with the setup of regulations, such as the free trade zone (FTZ) bordering, and collaboration to deal with oil spills produced by ships at the port harbors.

The prescriptive example includes:

- Agents: seaport authorities (individuals or associations).
- A representation of agents of at least two port decision makers: the agents are autonomous but act cooperatively.
- The visual and output functions of the system explained as a dynamic system able to re-create exemplary real contexts. Figure 9.2 shows the representation of a general dynamic process consisting of various classifications and variables performed in ports. Traffic lights indicate the behavior of the agent. A green light gives a free toll to those port agents accomplishing the restricted conditions. A yellow light calls for a caution within the conditions. Finally, a red light remarks on conditions for improvement that may be a matter of concern when the situation is to bring collaboration into play. Agents of interest among all the possibles are detected. An ideal in the sequence is the selection of a particular agent for a closer examination of the benefits it brings to integrate information, and to whom risk detection and mitigation might be realized. Key variables are visualized by rectangles in blue. Respective features or measures are visualized by rectangles in skintone. Arrows represent possible values for each feature. Ovals represent agents (individuals or associations), considering two homogeneous groups ('initiative' and 'passive' agents). Data flows may be cyclical, i.e., it is possible to follow closed and open loops when analyzing the various paths agents can take. The support for time variant workflows is provided by monitoring data series.
- In answering what is the knowledge/epistemological level that can describe what the agents need to know, an important issue in knowledge-based systems consists of the use of data supporting users in decision support for its formal modeling (Burstein et al. 2008). According to the International Maritime Organization (IMO 2011), PSC allows data for the inspection of foreign vessels in national ports in compliance with safety and security legal regulations. PSC is a well-known control mechanism regulated by the Memorandum of Understandings (MoUs) between regional/states agreements and their seaport authorities. An example of a data flow sequence uses reported cases of alleged inadequacies of reception facilities as decision-making variables for collaboration to deal with oil spills produced by ships at the port harbors and the FTZ bordering. See data descriptions in Table 9.1 for an example.
- The simulated ecosystem is dynamic and obeys ecological and normative principles given at the PSC context, which is a well-known mechanism of collaboration to deal with oil spills produced by ships at the port harbors and the FTZ bordering at several seaport jurisdictions.
- The types of support for the decision makers accomplish the set of values, queries, and scenarios that contribute to the identification of design choices targeted by seaport authorities.

9.6 FUTURE RESEARCH DIRECTIONS

It is desirable to find an open and freely available agent-based simulation tool, which could be extended and customized for specific situations. Scenarios need to

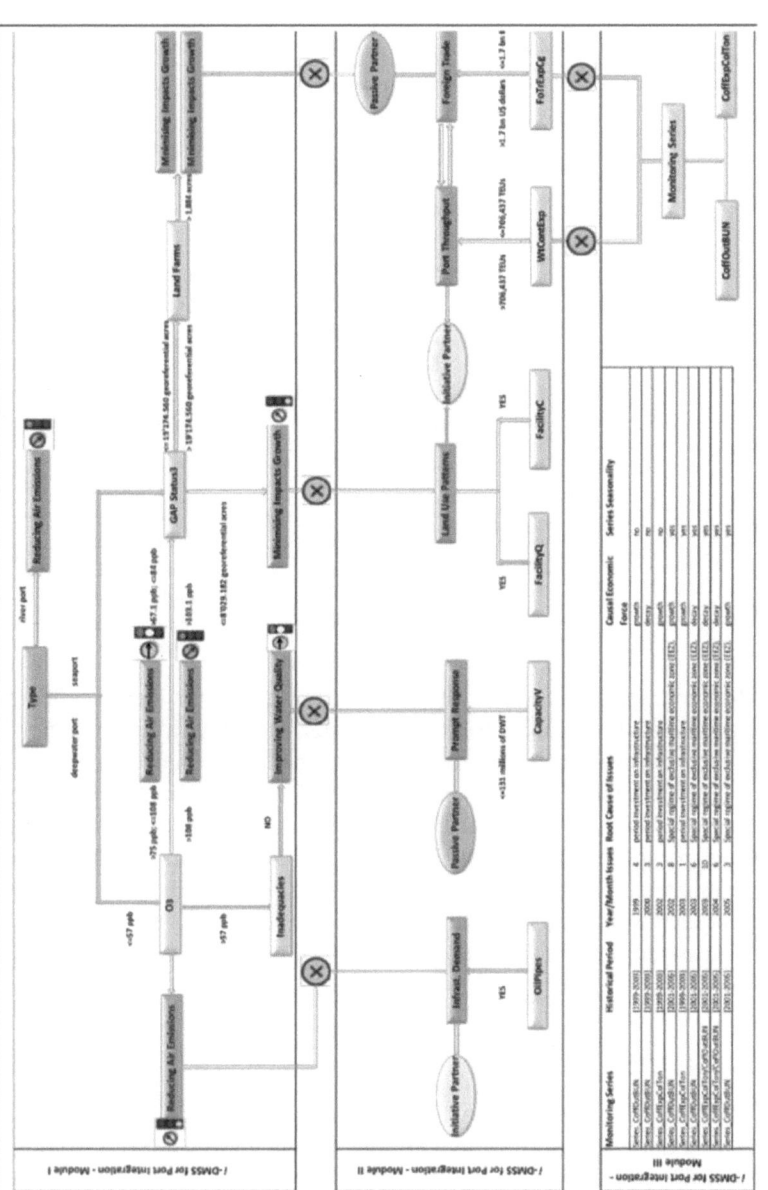

FIGURE 9.2 Visual and output functions of the agent-based system for seaport authorities' collaboration

Note: A color version of this figure can be found in the ebook.

Source: Halabi-Echeverry, 2016.

TABLE 9.1
Data Description at the PSC Context

Code (w's)	Name	Value	Source	Comments
w11	Inadequacies	Binominal {yes=1, no=2}	IMO International Maritime Organization https://gisis.imo. org/Public/PRF/ Defeault.aspx	Yes/no historical Encountered difficulties in discharging waste to reception facilities at the port Informed by IMO
w12	Facilities	Ordinal {>28, <=28}	IMO International Maritime Organization https://gisis.imo. org/Public/PRF/ Defeault.aspx	Updated number of port reception facilities (be they fixed, floating or mobile) in which final disposal of MARPOL residues/ wastes occurs in a manner that protects the environment, the health and safety of workers, and general population

be developed to encourage seaport authorities to utilize the tool for identification of decision-making choices. Therefore, a friendly user interface is highly recommended.

Among suitable tools, we have found the Evoplex application which comes with a user-friendly and intuitive GUI, plugins, and a variety of widgets. These allow for the creation and running of experiments to analyze (or visualize) the outputs, and it also allows changes on parameters to be done regularly, so the simulation solution can incorporate appropriate configuration and validation features (Cardinot et al. 2019).

Finally, a great challenge is to describe the formalism of the agent-based knowledge by means of modal logic so it is possible to develop propositional semantics for decision choices.

9.7   CONCLUSIONS

We state that two possible frameworks to support the agent's and multi-agents' interaction and communication protocols (i.e., at the knowledge level to support the dynamic behavior needed for the agents interactions) are: (1) IEEE-FIPA and, (2) BSPL.

Agent's knowledge represented semantically is a challenging universe of discourse that contributes to the identification of decision-making choices. In that sense, we propose exemplary real-world contexts in which the phenomena occur, such as the decision process for controlling, mitigating, and reducing risk emissions produced by ships and ports in collaboration with seaport partners, and taking into consideration the recommendations of the PSC mechanisms. An agent-based methodology contributes to the outcome of the decision process in this domain.

REFERENCES

Acciaro, M., T. Vanelslander, C. Sys, et al. 2014. "Environmental sustainability in seaports: A framework for successful innovation." *Maritime Policy & Management,* 41 (5): 480–500.

Benali, H. and N. B. Ben. 2011. "Towards a component-based framework for interoperability and composability in modelling and simulation." *Simulation,* 87 (1–2): 133–148.

Benenson, I. and P. M. Torrens. 2004. *Geosimulation: Automata-based modelling of urban phenomena.* Hoboken, NJ: Wiley.

Bennet, A. and D. Bennet. 2008. "The decision-making process in a complex situation." In *Handbook on Decision Support Systems 1,* edited by Frada Burstein and Clyde W. Holsapple. Berlin: Springer: 3–20.

Burstein, F., S. McKemmish, J. Fischer, et al. 2008. "A role for information portals as intelligent support systems: Breast cancer knowledge online experience." In *Intelligent Decision Making: An AI-based approach,* edited by G. Phillips-Wren, N. Ichalkaranje, and L. C. Jain. Berlin: Springer: 389–383.

Cardinot, M., C. O'Riordan, J. Griffith, et al. 2019. "Evoplex: A platform for agent-based modelling on networks." *SoftwareX,* 9: 199–204.

Cassaigne, N. and L. Lorimier. 2006. "A challenging future for i-DMSS." In *Intelligent Decision-making Support Systems,* edited by Jatinder N. D. Gupta, Guisseppi A. Forgionne, and Manuel Mora T. London: Springer: 401–422.

Chellas, B. F. 1980. *Modal Logic: An introduction.* Cambridge: Cambridge University Press.

Chi, L., E. Hartono, C. W. Holsapple, et al. 2008. "Organizational decision support systems: Parameters and benefits." In *Handbook on Decision Support Systems 1,* edited by Frada Burstein and Clyde W. Holsapple. Berlin: Springer: 433–468.

DeTombe, D. 2001. "Compram, a method for handling complex societal problems." *European Journal of Operational Research,* 128 (2): 266–281.

DeTombe, D. 2002. "Complex societal problems in operational research." *European Journal of Operational Research,* 140 (2): 232–240.

DeTombe, D. 2013a. "The actors of the credit crisis reflected by the Compram methodology." *Central European Journal of Operations Research,* 21 (1): 1–29.

DeTombe, D. 2013b. "How to handle societal complexity." In *Selected Topics in Nonlinear Dynamics and Theoretical Electrical Engineering,* edited by K. Kyamakya, W. A. Halang, W. Mathis, et al. Berlin: Springer: 227–244.

Ferrari G., U. Montanari, and E. Tuosto. 2003. "Graph-based models of internetworking systems." In *Formal Methods at the Crossroads. From Panacea to Foundational Support,* edited by B.K. Aichernig and T. Maibaum. Berlin: Springer: 242–266.

FIPA (Foundation for Intelligent Physical Agents). 2002. "FIPA Abstract Architecture Specification." www.fipa.org/specs/fipa00001/SC00001L.html.

Gachet, A. and P. Haettenschwiler. 2006. "Development processes of intelligent decision-making support systems: Review and perspective." In *Intelligent Decision-making Support Systems,* edited by Jatinder N. D. Gupta, Guisseppi A. Forgionne, and Manuel Mora T. London: Springer: 97–121.

Georgakaki, A., R. A. Coffey, G. Lock, et al. 2005. "Transport and Environment Database System (TRENDS): Maritime air pollutant emission modelling." *Atmospheric Environment,* 39: 2357–2365.

Giannouli, M., Z. Samaras, M. Keller, et al. 2006. "Development of a database system for the calculation of indicators of environmental pressure caused by transport." *Science of the Total Environment*, 357: 247–270.

Gruber, T. R. 1995. "Toward principles for the design of ontologies used for knowledge sharing." *International Journal of Human-Computer Studies*, 43 (5): 907–928.

Halabi-Echeverry, A. X. 2016. "Computational intelligence for development of strategic decision making in port informational integration." Macquarie University, Sydney, Australia.

Han, J., Y. Hayashi, X. Cao, et al. 2009. "Application of an integrated system dynamics and cellular automata model for urban growth assessment: A case study of Shanghai, China." *Landscape and Urban Planning*, 91 (3): 133–141.

IMO (International Maritime Organization). 2011. "Port State Control." www.imo.org/en/OurWork/MSAS/Pages/PortStateControl.aspx.

Lee, S. W., D. W. Song, and C. Ducruet. 2008. "A tale of Asia's world ports: The spatial evolution in global hub port cities." *Geoforum*, 39: 372–385.

Luan, W. X., H. Chen, and Y. W. Wang. 2010. "Simulating mechanism of interaction between ports and cities based on system dynamics: A case of Dalian, China." *Chinese Geographical Science*, 20: 398–405.

Macal, C. and Michael North. 2010. "Tutorial on agent-based modelling and simulation." *Journal of Simulation*, 4 (3): 151–162.

Murty, K. G., J. Y. Liu, Y. W. Wan, et al. 2005. "A decision support system for operations in a container terminal." *Decision Support Systems*, 39 (3): 309–332.

Negash, S. and P. Gray. 2008. "Business intelligence." In *Handbook on Decision Support Systems 2*, edited by Frada Burstein and Clyde W. Holsapple. Berlin: Springer: 175–193.

Noff, S. 2009. "Automation: What it means to us around the world." In *Springer Handbook of Automation*, edited by S. Y. Nof. Berlin: Springer: 13–52.

Pomerol, J.-C. and F. Adam. 2006. "On the legacy of Herbert Simon and his contribution to decision-making support systems and Artificial Intelligence." In *Intelligent Decision-making Support Systems*, edited by Jatinder N. D. Gupta, Guisseppi A. Forgionne, and Manuel Mora T. London: Springer: 25–43.

Russell, S. and P. Norvig. 1995. *Artificial Intelligence: A modern approach*. Englewood Cliffs, NJ: Prentice Hall.

Russell, S., P. Norvig, and E. Davis. 2010. *Artificial Intelligence: A modern approach*. 3rd edition. Englewood Cliffs, NJ: Prentice Hall.

Singh, M. P. 2011. "BSPL, the Blindingly Simple Protocol Language." North Caroline State University. www.csc2.ncsu.edu/faculty/mpsingh/papers/mas/AAMAS-11-IBIOP.pdf.

Sirikijpanichkul, A., K. Vandam, L. Ferreira, et al. 2007. "Optimizing the location of intermodal freight hubs: An overview of the agent-based modelling approach." *Journal of Transportation Systems Engineering and Information Technology*, 7 (4): 71–81.

Tongzon, J., Y. T. Chang, and S. Y. Lee. 2009. "How supply chain oriented is the port sector?" *International Journal of Production Economics*, 122: 21–34.

Wiek, A. and A. Walter. 2009. "A transdisciplinary approach for formalized integrated planning and decision making in complex systems." *European Journal of Operational Research*, 197: 360–370.

Wooldridge, M., N. R. Jennings, and D. Kinny. 2000. "The Gaia methodology for agent-oriented analysis and design." *Autonomous Agents and Multiagent Systems*, 3 (3): 285–312.

Yeo, G. T., M. Roe, and J. Dinwoodie. 2008. "Evaluating the competitiveness of container ports in Korea and China." *Transportation Research Part A: Policy and Practice*, 42: 910–921.

10 Simulation and Reinforcement Learning Framework to Find Scheduling Policies in Manufacturing

Edgar Gutierrez, Nicolas Clavijo-Buritica, and Luis Rabelo

10.1 INTRODUCTION

Nowadays, integration between decision levels (strategic, tactical, and operational) is a research interest of academia and industry. An essential process in this endeavor has been more notable in the execution and operations levels. The central application area has been in the manufacturing industry (Chu and You 2015). Measuring the impact of the operational decisions in manufacturing is one of the challenges many companies encounter. Holding better methodologies to support the coordination between execution and the other levels helps to benefit the organization. Usually, decisions follow a policy, a previous plan, and a schedule (Rabelo, Jones, and Yih 1994). Due to unforeseen events, decisions may need to be made in a short time under an uncertain environment in the execution phase. This work presents a framework to support decisions under scheduling planning. This framework is composed of a system with a reinforcement learning algorithm with a discrete simulation model to create a feedback learning loop. The goal of the framework is to find policies that maximize an objective scheduling function in job shop scheduling. The policies determine what actions to take to get the best schedule process.

For this framework, one of the algorithms used mainly by the Machine Learning community is utilized. Q-Learning is used for machine scheduling and selection of promising dispatching rules which will define the performance level of the manufacturing system. The procedure considers the generation of pairs of queue states and dispatching rules. The logic of the procedure suggests the assignment of Q-values to the queue states for each dispatch rule and then a process of updating the Q-values based on the performance of the manufacturing system is deployed.

Neural networks are used as learning functions. Backpropagation is utilized as the utility network. In the context of the scheduling problem, the conventional attributes such as arrival times, lead times, processing times, and delivery dates are determined stochastically. At the same time, these attributes are the ones that allow defining the attributes of the tasks in a multiple neural network of a single output. While the output of each neural network is a Q-value that is assigned to each dispatch rule, the input is represented by the queue state (the state based on the attributes of the tasks). By means of a stochastic process (in this case a discrete simulation), one of the possible Q-values associated with the actions is selected and then evaluated in the manufacturing system. The performance of the system (work-in-process inventory, machine utilization, or mean tardiness) is evaluated by completing the action. When the system's performance metrics show an unfavorable result with the selected action, then it is penalized. Otherwise, it is rewarded. Thus, when the process is repeated, the weights that accompany the neural network are adjusted and updated, defining lower Q-values for the outputs that present unfavorable results and higher values for those cases that represent improvement in the performance function. This mechanism increases the chance of selecting actions that will improve the performance of the manufacturing system. After multiple repetitions of the procedure, the Q-values converge optimally.

Qu, Wang, and Shivani (2016) researched reinforcement learning concerning manufacturing real-time processing capabilities. The focus was on handling multiple manufacturing processes, complex job requirements set by orders, and how the construction of reinforcement learning is capable of handling multiple orders and processes with optimal results. It was discovered that using the reinforcement learning method resulted in better outcomes of complex order processes than the traditional process of dispatching orders. In addition, research by Sharabi, Amin, and Mahootchi (2017) focused on the relation of reinforcement learning used to enhance the methods utilized in scheduling in the manufacturing process. The reinforcement learning method successfully used a reward-based learning process and projected high performance in the controlled environment (Qu, Wang, and Shivani 2016).

10.2 PLANNING AND SCHEDULING FOR PRODUCTION SYSTEMS

Preceding literature review in this topic discusses theories, models, applications, methods, and methodologies in which the majority are related to production scheduling and routing problems. For instance, Harjunkoski and colleagues (2014) discuss production scheduling problems focusing on the models' strengths and weaknesses. An integration between control (execution level) and scheduling (operative level) has been reported as the key for successful operational processes (Baldea and Harjunkoski 2014; Harjunkoski, Nyström, and Horch 2009). One of the challenges many companies face is the lack of design integration for operational decisions (Shah 2005).

Hierarchical structures have been proposed mainly for production in manufacturing systems. Usually, different decision levels are organized depending on the impact, purposes, and planning horizon, based on the hierarchical model proposed by the International Society of Automation (ISA). The upper layer deals with strategy and tactical decisions, where decisions should be taken for long and medium terms. It is followed by and connected with the manufacturing operations, which set the activities to meet the final product under medium and short time frames and is mainly considered under operative decision level in a supply chain. Finally, at the sensing and execution levels, where automatic control systems, dashboards, and communication systems support the operation, data integration and information flow are allowed. Some examples of such systems are listed as Enterprise Resource Planning (ERP), Manufacturing Execution Systems (MES), and Supervisory Control and Data Acquisition (SCADA) (Chu and You 2015). Those systems and the interaction between the different layers allow for collecting and analyzing information quickly. Going from the execution level with the actions' results to the management level and transferring instructions downward to carry out the actions. Having better methodologies to support coordination between the execution levels helps reach benefits for the organization. Decisions are addressed from decision makers in strategic levels to the operative ones, passing through the tactical levels.

10.2.1 Production Scheduling Environments

The scheduling in manufacturing facilities organized by tasks (job shop scheduling) is perhaps one of the most complex due to this environment's particularities. The scheduling function can be a simple or very complex task depending on the production approach and the workshop's productive configuration (Sahinoglu 1993). Figure 10.1 illustrates a schedule for a job shop with eight machines.

What complicates the scheduling function in a job shop environment is the variety of orders that must be processed, each with a different processing path and in small batches that do not allow continuity and require a high number of changes. In the 1970s, with the rise of complexity theory (Lenstra and Rinnooy Kan 1978), programmers classified earlier problems as NP-Hard (Non-deterministic Polynomial-time Hard) (Brucker, Hilbig, and Hurink 1999).

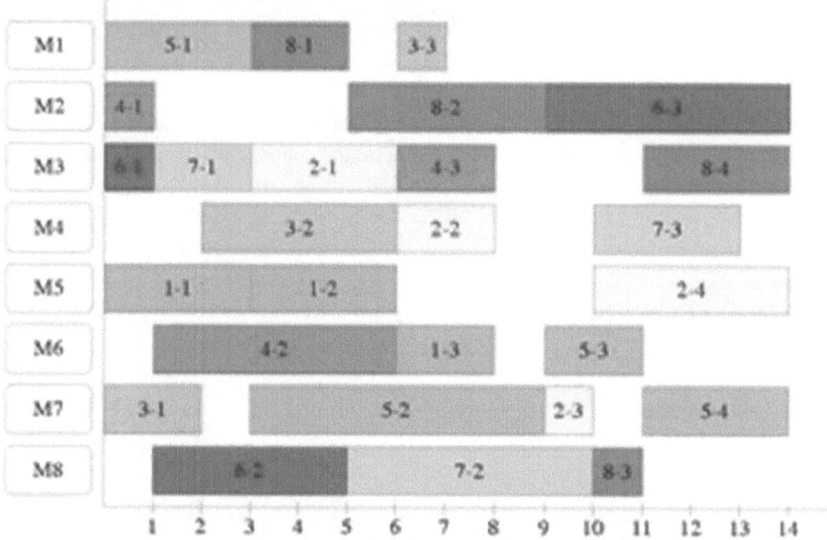

FIGURE 10.1 Example of a schedule for a job shop problem with eight machines

10.2.2 Integration of Operational and Executional Level

Several researchers have studied schedules that include a control stage (Castro et al. 2011; Baldea and Harjunkoski 2014; Harjunkoski, Nyström, and Horch 2009). Other applications based on hybrid models (HM) and Artificial Intelligence (AI) have been used to attend scheduling and control problems (Rabelo, Bhide, and Gutierrez 2018; Castro et al. 2011).

In addition, Ivanov, Dolgui, and Sokolov (2012) studied applications and extensions of control theory (CT). The authors describe essential issues and perspectives that delineate dynamics in supply chains while identifying CT, related to production, logistics, and SCM from 1960 to 2011. For decades, the use of artificial neural networks (ANNs) has been successfully applied to solve various problems. Sabuncuoglu and Gurgun (1996) propose a new neural network approach to solve minimum tardiness for single machine scheduling problems and minimum make-span for the job shop scheduling problem. In detail, they mainly used mathematical programming and AI for problems' execution. However, uncertainty in planning, scheduling, and control are the primary concerns. Therefore, some works were found to attend real-time optimization problems while others focused more on operations and execution problems such as scheduling and re-scheduling, and routing and re-routing.

10.3   LEARNING SCHEDULING

For the production scheduling process in a manufacturing environment, we need to evaluate what the *decision variables* should be for the scheduling. Questions such as: What sequence should the systems follow? What should be the order in which the jobs are carried out? How should the calendar be defined? All this to define the assignment of jobs based on time and environment. In this phase, the importance of how to learn and choose the best policy is revealed. In this way, the *programming policies* can be defined later.

This research work presents a methodology to use Machine Learning techniques to answer these questions and determine the sequences in which the work must be carried out, while taking into account the particular characteristics of each system. Other variables are:

- Completion time (C_{ij})
- Lateness (L_j)
- Tardiness (T_j)
- Make-span (M_j)

An agent-based model (ABM) models agents' actions and interactions to evaluate their effects on the system (see Figure 10.2).

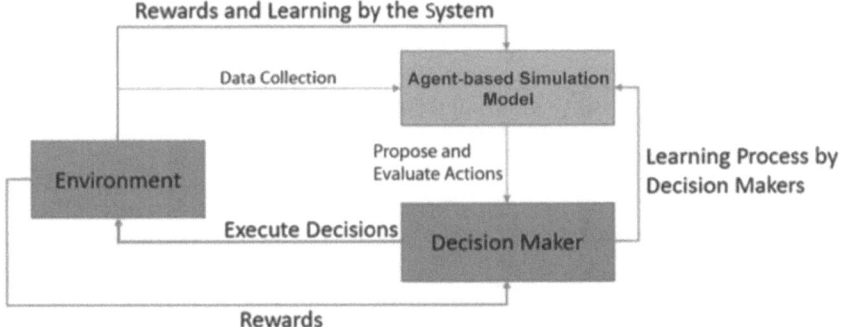

FIGURE 10.2 Agent-based system to model actions and evaluations with specific rewards

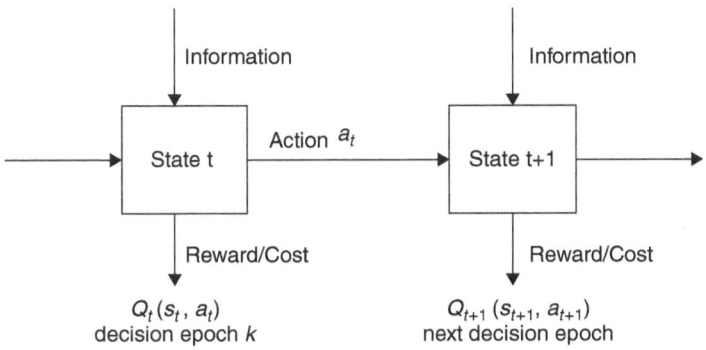

FIGURE 10.3 Sequence of decisions in an MDP

10.3.1 MARKOV DECISION PROCESS

Four main parts represent the Markov Decision Process (MDP):

1. A set of state spaces {s1 ⋯ SN}: Which contain the information to make decisions and each epoch k.
2. A set of actions {a1 ⋯ aM}: An action should be selected, and each decision epoch k.
3. A set of rewards {r1 ⋯ rN}: (One for each state), such that R (sk, ak) represents the reward obtained by the agent in state sk after taking action ak. The reward is calculated as a contribution and comes from the calculation of the performance indicators. RL is expected to learn the best action to obtain the best reward to improve the whole system.
4. A set of policies: Policy π is a sequence of decisions. ∏ represents the set of all possible policies (e.g., Lookahead, value function approximation policies). Figure 10.3 displays the different relationships between actions and rewards according to the states.

Objective: Objectives are defined as action-value functions that lead to choosing the best policy. Discount factor λ ($0 < \lambda < 1$) is necessary for the objective function.

A transition probability function P is a Markovian transition model where $P(x_j \mid x_i, k)$ represents the probability of going from state x_i to state x_j with action α

$$P_{ij}^k = Prob(Next = j \mid From = i \text{ and using action } k) \tag{1}$$

The cost or value for each state depends on the future rewards (feedback). The total value is represented by $Q_t (S_t, a_t)$ of the actions taken in state t, and is the sum of the immediate reward and the approximation of the value of the next state:

$$Q_{t+1}\left(S_t, \alpha_t\right) = Q_t\left(S_t, \alpha_t\right) + \alpha\left[r_{t+1} + \lambda max Q_t\left(S_{t+1}, \alpha_{t+1}\right)\right] \tag{2}$$

The learning rate and discount factor used in the simulations are represented by α and λ, respectively.

For our example, with the reinforcement learning (RL) concept, we propose to solve the MDP. One of the main characteristics of the RL method is the use of rewards. The system learns what to do through time and can map situations into actions to maximize the total reward signal. RL adjusts the policies due to observations and reinforces the right actions relative to the wrong actions. Rewards represent the desired goals, which are calculated with our performance indicators. By maximizing these indicators, the algorithm will improve the system to reach the goals. These indicators are continuously calculated due to the learning interactions of different "agents" and the environment. Uncertainty comes from two primary sources: (1) the incorporation of customer demand, and (2) a real-time flow of information from customers and drivers. The main objective is to find the best action for each state (policy) that accomplishes as much reward as possible.

10.3.2 Learning and Scheduling of Jobs Framework

RL is a technique useful in solving control optimization problems (Gosavi 2009). By control optimization, we mean recognizing the best action in every state visited by the system to optimize some objective function, e.g., the average reward per unit time and the total discounted reward over a given time horizon. Sahinoglu (1993) discusses the situation that arises when you want to schedule independent single-operation jobs available to process during normal hours and overtime. The system defines the arrival behavior, the process plan, the distribution of processing time, and the dependencies.

As is known in machine programming problems, the total work setup times will be dependent on their sequence. When the machine is available, it does not idle while the job is on hold. Once a job begins processing, it completes without interruption. In these cases, pre-emption (interrupting the process to process other jobs) is generally not allowed. Figure 10.3 depicts a framework for integrating and utilizing scheduling techniques with the Q-Learning paradigm (Watkins and Dayan 1992), which leads to achieving desired performance levels. A scheme was developed to

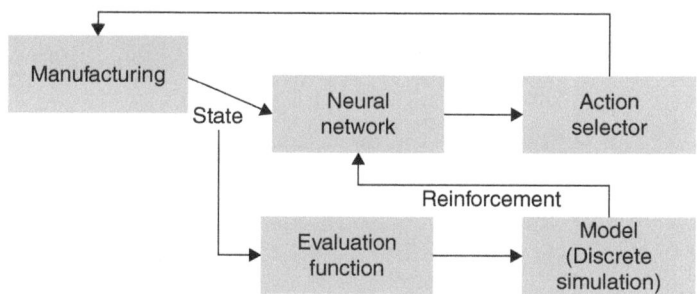

FIGURE 10.4 Framework of Q-Learning with the RL model

achieve this integration. The problem of applying Q-Learning in scheduling problems occurs in the reinforcement stage. The reinforcement mechanism works based on the differences between the previous and current values. There could be some cases in which stochastic action selectors choose the action with the system's best effect. Indeed, the difference in the system performance's current and previous values might require punishments for the system.

10.4 ILLUSTRATIVE EXAMPLE

Q-Learning is used for modeling evaluation function. For a provided state, where the utility is maximum, the system chooses the action. The model assigns values to action–state pairs. Neural networks should then be used to renovate the Q-values. The stochastic action uses a probability distribution to determine the Q-values. The model's part is to evaluate each action's effect via simulation and to inform the network whether there is a penalty or a reward. Actions are simulated independently, and the earlier and present values of the system performance are calculated for each action. Agreeing to the results of the simulation, actions are ranked regarding the difference. The type of reinforcement is marked based on the relative difference of previous and current performance criteria. This procedure is applied to the program in each iteration. The main steps are described in Figure 10.5.

The "agent" decides an action, gains feedback (good or bad) for that action, and uses the feedback to update its record. In its record, the agent keeps a so-called Q-factor for every state–action pair. When the feedback for selecting an action in a state is positive, the associated Q-factor's value is increased, while if the feedback is negative, the value is decreased. The feedback consists of the immediate revenue or reward plus the value of the next state.

The agent uses the RL algorithm to update its knowledge base, becomes smarter in the process, and then selects a better action. In this illustrative example, we use the same procedure as described in Section 10.1. Figure 10.6 represents the sub-routine in which the rewarded action is evaluated in the manufacturing environment. For this case, a discrete simulation model is used for the single machine scheduling problem, instead of making changes to a real system. Once the simulation model is run, the performance variable Cmax (make-span) is evaluated for each rule studied.

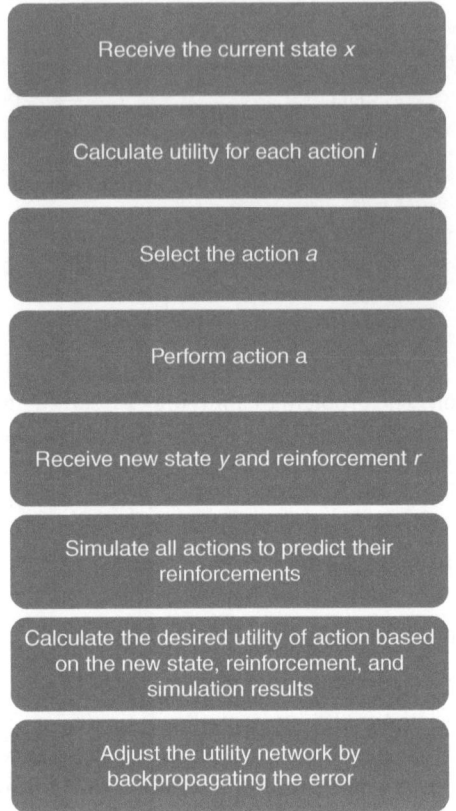

FIGURE 10.5   Simulation actions and reinforcements

Once the control receives queue status from the manufacturing environment (step1) and translates queue status into input format (step 2), the next step is to use the neural networks to calculate Q-values for each action (step3), as shown in Table 10.1.

The following is a short description of the next steps:

Step 4: Stochastic action selector chooses an action: Action = 4 (Perform LIFO).
Step 5: The manufacturing system performs the action and starts processing the job.
Step 6: The manufacturing system sends the performance criterion.
Step 7: The controller compares the previous and current values of the performance criterion.
Step 8: The controller receives the new queue status.
Step 9: The controller translates new queue status into the input format.
Step 10: Neural networks calculate Q-values for each action.
Step 11: The controller takes the maximum Q-value from the new state's value.
Step 12: The controller simulates all the actions and ranks them.
Step 13: The controller defines the type of reinforcement.

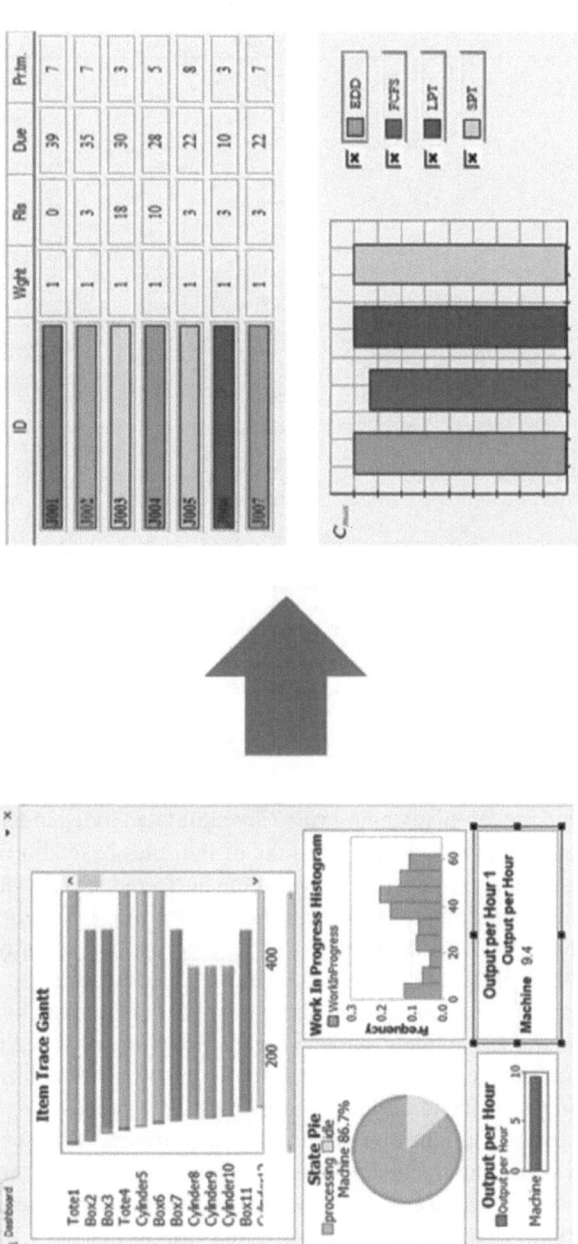

FIGURE 10.6   Gantt chart, showing scheduling policies and the respective updates

TABLE 10.1
List of Dispatching Rules and Corresponding Q-value

Rule	Q-value
SPT	0.519
LPT	0.462
FIFO	0.424
LIFO	0.515
CR	0.537
MSlack	0.446
mSlack	0.464
EDD	0.487
LDD	0.506
SST	0.553
LST	0.426
SPST	0.455
LPST	0.438
SSlack	0.462
Slack/RT	0.471

Step 14: The controller reinforces the network by using learning parameters and maximum Q-value for the successor slate.

Step 15: The process continues from step 4.

10.5   CONCLUSIONS

In Q-Learning with a model, dispatching rules are simulated independently, and each simulation result should be recorded. The ranks of the rules based on the simulation results indicate each dispatching rule's effect on the performance criterion. A discrete model is used to evaluate all actions through a simulator to provide the system with information on what kind of reinforcement should be applied. Finally, actions are simulated after the completion of the selected action. Deep Learning in the manufacturing process has two factors that can be distinguished: (1) the systems for representation of knowledge that use models, and (2) the methodology of scheduling that uses algorithms. Our work entails using a hybrid modeling approach that has proven good results in forecasting issues and adjusting to using a new process in other research work. Also, our research is unique due to using Deep Learning, which can make using the real-time system in manufacturing operations management and manufacturing execution systems simpler to users in the manufacturing process. Therefore, we feel our research is distinguishable from other work in a similar field.

ACKNOWLEDGMENTS

This material is partially based upon work supported by the National Science Foundation under award no. 2012228.

REFERENCES

Baldea, Michael and Iiro Harjunkoski. 2014. "Integrated production scheduling and process control: A systematic review." *Computers & Chemical Engineering,* 71 (December): 377–390.

Brucker, Peter, Thomas Hilbig, and Johann Hurink. 1999. "A branch and bound algorithm for a single-machine scheduling problem with positive and negative time-lags." *Proceedings of the Third International Conference on Graphs and Optimization GO-III,* 94 (1): 77–99.

Castro, Pedro M., Adrián M. Aguirre, Luis J. Zeballos, et al. 2011. "Hybrid mathematical programming discrete-event simulation approach for large-scale scheduling problems." *Industrial & Engineering Chemistry Research,* 50 (18): 10665–10680.

Chu, Yunfei and Fengqi You. 2015. "Model-based integration of control and operations: Overview, challenges, advances, and opportunities." *Computers & Chemical Engineering,* 83: 2–20.

Gosavi, Abhijit. 2009. "Reinforcement learning: A tutorial survey and recent advances." *INFORMS Journal on Computing,* 21 (May): 178–192.

Harjunkoski, Iiro, Christos T. Maravelias, Peter Bongers, et al. 2014. "Scope for industrial applications of production scheduling models and solution methods." *Computers & Chemical Engineering,* 62 (March): 161–193.

Harjunkoski, Iiro, Rasmus Nyström, and Alexander Horch. 2009. "Integration of scheduling and control—Theory or practice?" *Computers & Chemical Engineering,* 33 (12): 1909–1918.

Ivanov, Dmitry, Alexandre Dolgui, and Boris Sokolov. 2012. "Applicability of optimal control theory to adaptive supply chain planning and scheduling." *Annual Reviews in Control,* 36 (1): 73–84.

Lenstra, J. K. and A. H. G. Rinnooy Kan. 1978. "Complexity of scheduling under precedence constraints." *Operations Research,* 26 (1): 22–35.

Qu, Shuhui, J. Wang, and G. Shivani. 2016. "Learning adaptive dispatching rules for a manufacturing process system by using reinforcement learning approach." *2016 IEEE 21st International Conference on Emerging Technologies and Factory Automation (ETFA):* 1–8.

Rabelo, L. C., A. Jones, and Y. Yih. 1994. "Development of a real-time learning scheduler using reinforcement learning concepts." *Proceedings of 1994 9th IEEE International Symposium on Intelligent Control:* 291–296.

Rabelo, Luis, Sayli Bhide, and Edgar Gutierrez. 2018. *Artificial Intelligence: Advances in research and applications.* New York: NOVA Science.

Sabuncuoglu, Ihsan and Burckaan Gurgun. 1996. "A neural network model for scheduling problems." *European Journal of Operational Research,* 93 (2): 288–299.

Sahinoglu, Mehmet Murat. 1993. "Development of a Real-Time Learning Scheduler Using Adaptive Critics Concepts." Ohio University. https://etd.ohiolink.edu/apexprod/rws_olink/r/1501/10?p10_etd_subid=59638&clear=10.

Shah, Nilay. 2005. "Process industry supply chains: Advances and challenges." *Computers & Chemical Engineering,* 29 (6): 1225–1235.

Sharabi, Jamal, Mohammad Amin, and Masoud Mahootchi. 2017. "A reinforcement learning approach to parameter estimation in dynamic job shop scheduling." *Computers & Industrial Engineering,* 110: 75–82.

Watkins, Christopher J. C. H. and P. Dayan. 1992. "Technical note: Q-learning." *Machine Learning,* 8: 279–292.

11 An Advanced Analytical Proposal for Sales and Operations Planning

Julio A. Padilla

11.1 INTRODUCTION

Sales and operations planning, known by its acronym S&OP, is a process of business coordination through a balance between demand management in the market and supply chain possibilities in search of a business plan that responds to the highest profitability within the limitations and strategy of the company. The achievements of this objective are highly questioned, but even accomplishing them, they are far from what a true business integration requires. An important point that confuses many companies is not to consider S&OP as an integration process that must have inputs and outputs that generate both the development of a business plan and the guidelines for its execution. But this process must have clearly defined stages, which should be led and executed by the appropriate managers. Most large and medium-sized companies have implemented this process with the indicated limitations. Fewer companies use software for the acquisition and integration of data relevant to the process, and even less are those that employ advanced analytics—that is, optimization models integrated with descriptive and forecasting models to support the important decisions that must be taken during the whole process. S&OP and all supply chain management (SCM) topics must move beyond functional improvement to deliver significant strategic and competitive advantage to companies, including sales and marketing decisions.

Nowadays, it is more important than ever to develop analytical capabilities to build robust supports for forecasting and decision making in SCM, such as those required by the S&OP process. Commercial companies selling software for SCM accept this importance and have been offering support to S&OP, but in most cases with only descriptive models. The few that do offer advanced analytics in their supports keep the corresponding design secret. It is time for academics to contribute to this regard and begin to propose clear and transparent solutions that form mature knowledge and inspire better commercial solutions.

This chapter presents an integrated support proposal for S&OP. On the supply management side, procurement, production, and distribution processes with their respective capabilities and costs are included. Inventories of materials in plants, of finished products in plants and distributors, their corresponding storage capacities,

DOI: 10.1201/9781003137993-12

and their costs, are also considered in the integration. On the side of demand management, sales potentials are included in terms of forecasting as function of average prices—that is, the relation of revenue and sales with respect to prices, and the possibility of defining a marketing strategy. The resulting model has been linearized to obtain a mixed-binary programming model as a tool capable of solving the complex problems that result in real S&OP within prudential times.

The proposal is for mass-consumer B2C companies where usually there are lots of discounts and promotions. With traditional forecasting techniques, few companies can forecast effectively in such a promotion-heavy environment. On the other hand, many companies have adopted growth through innovation strategies, incrementing the number of active items much more compared to growth in sales. As a result, sales per item have dropped, complicating the forecast process even more. A solution to this problem is the use of Machine Learning (ML) techniques to increase forecast reliability and to simplify the process. Today, enterprises are attaining major improvements in forecast error rates and demand planning productivity using ML.

This chapter presents a Python code to apply one of these ML techniques in an appropriate form to feed the proposed decision-making model. Most current use of ML is in near-future forecast, but here it is used in mid-term monthly forecast. In addition, the requirement will be to have the relationship of sales and income based on prices. Price elasticity seems to be the solution but very few companies have developed this knowledge for their products. A few years ago, price elasticity promised to profitably shape demand and its optimization to greatly improve S&OP decisions. Over time, the discussion of price optimization in SCM is hard to find in academic literature. Besides, businesses have been less quick to implement analytics in SCM than in other areas of operation. What we propose is an ML method to achieve better estimates and understandings of a finished product sales when it is affected by a certain average price.

11.2 BACKGROUND

The S&OP process started in the 1980s and has undergone great evolution throughout its existence (Simchi-Levi 2008). In the beginning, it focused on demand analysis and planning. Later, it evolved into monthly meetings where companies compared the demand forecast with supply capacity restrictions, seeking to identify feasible execution plans. With the new approaches of total integration between demand and supply, S&OP acquires greater importance, but at the same time its inefficiencies are clearly denoted by Moon (2013) and Padilla (2014). The necessary evolution of S&OP and its different stages of maturity are excellently presented by Cecere and Chase (2013), who highlight the importance of an efficient and iterative S&OP process for business growth and profitability. With similar approaches, Burrows (2012) presents a revolutionary model for S&OP in the new economy on demand. He emphasizes the problem that most companies have immature S&OP processes, focused only on supply, and hence the need to use analytics to correctly incorporate the demand side.

Simchi-Levi (2010) presents the importance of the concept of flexibility in SCM and the need for S&OP to optimize decisions on various production sources considering changes in demand, costs, and business restrictions. In his excellent papers, Shapiro (2010a; 2010b) shows that companies are implementing S&OP with descriptive models and do not use advanced analytics. It highlights the importance of achieving mixed-integer programming models that allow iteratively analyzing complex business systems. He expresses that the link between business intelligence and advanced analytics is the critical technological development necessary for the new generation of S&OP.

Recent papers continue to emphasize the need to use advanced analytics, as in Mišić and Perakis (2020), who promote it in supply chain and price decisions in revenue management, but without integrating the two environments. They reference recent advances of ML to obtain high-quality predictive models for high-dimensional data and of optimization methodologies to increase application in operations management. Availability of high-quality software for estimating ML models and for solving large-scale linear mixed-integer optimization problems are highlighted. Like this chapter, they also discuss price-promotion planning—that is, when and how to promote each item over a time horizon.

Bertsimas and Kallus (2020) present a framework that combine ML and operations research (OR). Like them, this chapter considers these models in relation to the price as a decision variable that affects uncertainty, and that its casual effects are unknown, making necessary the development of demand curves derived from data. They emphasize that applications of ML generate predictions of quantities that are of interest to OR, but it is not clear how to go from a good prediction to a good decision. In the application aspect, they work with a shipment planning with pricing, finding that prediction using Random Forest performs the best.

Ferreira, Lee, and Simchi-Levi (2016) work with an online retailer using data to optimize pricing decisions daily. They use ML techniques to predict demands and develop an algorithm to solve the subsequent multi-product price optimization incorporating price effects. They found that Regression Trees outperformed other regression methods. This work is an excellent example of how to combine ML and OR techniques into a pricing decision support tool.

In applications directly related to S&OP, Taskin and colleagues (2015) present a mathematical programming model to support the S&OP process in a television production company, minimizing costs and not including pricing decisions. In an academic–business relationship, Scavarda and colleagues (2017) present a case method for the implementation of S&OP. An interesting application of diffuse logic in environments with subjectivity and uncertainty to determine the level of maturity of S&OP is presented by Pedroso and colleagues (2017). A centralized supply chain with products that show varying demand against price is presented in Kaplan and colleagues (2011). They formulate a non-linear mathematical programming model that requires approximation and reformulation to yield a solvable model.

Commercially, Logility, Inc (2018) emphasizes that developing robust analytical capabilities for supply chains is more important than ever, because it allows better business decisions that result in optimized operating outcomes and greater sensitivity to customer needs. E2open (2019) annually presents an interesting study about forecasting, inventory, and supply chain performance in North America. It draws attention to an indisputable truth; supply chain must go beyond delivering products and become an engine for profitability. Over the last nine years, the growth in active items outpaced the rise in sales by a factor of more than two, adding complexity and costs to operations. E2open criticizes this pointless innovation, since an analysis of product introduction data reveals that only one in a thousand new items becomes a top mover. It emphasizes the importance of: control over forecasting, the forecast error, and the bias as a measure of the quality of a S&OP process. Positive bias is usually an indication of incentive conflicts in the S&OP process, especially for organizations where the sales managers hold a disproportionate influence. E2open concludes that accurate forecasting is strategic since every decision by management is ultimately based on a forecast. In Europe, a recent survey by Lofvers (2020) shows that supply chain software implementations will be mainly focused on S&OP and end-to-end supply chain visibility, depending on the level of company supply chain maturity.

On the demand side, Senthilnathan (2016) performs a theoretical analysis of the impact of price elasticity on the demand and the revenue of a company. Meissner and Strauss (2009) dynamically analyze the price decision in the well-known revenue management models, approximating their dynamic model through a mixed-linear program with success. The need and lack of work in the price interaction and the performance of the supply chain is highlighted by Fleishmann, Hall, and Pyke (2011), who propose a dynamic finite horizon model that seeks to understand the nature of optimal prices and promotions.

11.3   PROCEDURES

The purpose of the framework used in this chapter's research is to optimize the marginal contribution of a company through pricing decisions (discounts or promotions) and utilization of supply capacity; typical decisions of the S&OP process. The objective is to decide at what monthly average prices to market each of the company's products and simultaneously to decide how to allocate the capacity of the company's supply resources among those products to maximize the marginal contribution generated by the sales of all products, given the accomplishment of a market strategy.

As in every prescriptive decision, it is necessary to have the best forecasting of the future demand. In our case this is more challenging because the proposed system needs the estimated demands and their relation to the average price at which the company will be selling. This research does not pretend to work in the theoretical development of price elasticity of demand. Remembering that the scope of the research is mass-consumer B2C products, the hypothesis is that those relations are in relatively easy-to-find data. This research is talking about the billing figures of the company. Getting the monthly sales volume and the corresponding billing figure for a product, it is possible to obtain the average price. Mass-consumer products are usually affected by promotions and discounts so you will be finding different values of the average price in each month. This is an example of how SCM driven by forecasting must be changed by new technologies collecting data inside the company or outside (as in social media). Traditional regression and time series analysis are now competing with Artificial Intelligence subsets such as Random Forest in demand understanding and forecasting. The advantages of these new technologies and how to use them will be explained later in this chapter.

11.3.1 PREDICTING SALES

What we need to get is the monthly forecast for the company sales in each item and its relation to the average price. Without a quality forecast it is impossible to make a S&OP process more efficient and effective. By using ML and with the real data of a company that sells mass-consumer food products, it will be possible to develop a procedure to make an efficient forecast. The main data for the forecasts is the monthly sales figures, and fortunately they were ready to use, but we also need the corresponding billing figures, and it required a little effort to get them. Dividing the billing by the sales, we get the average price for each month and it will be one of the variables to use in the ML procedure.

To determine in which items it is possible to obtain a good forecast, it was necessary to use a typical segmentation looking for forecastability. The two drivers used to form a matrix are: average monthly volume (AV) and coefficient of variation (CV). To compare these drivers among all the products of the company, the monthly sales had to be normalized to the same unit. For this company that unit was kilos. Additionally, with the help of the marketers, all the sales levels included in the segmentation were classified with two behaviors: marketing and statistical. Marketing behavior means that there is an understanding of the corresponding customers cluster and statistical behavior means that there is enough volume in the monthly sales. Some products work with only one figure for the complete sales, but most of them were divided by market zones. Figure 11.1 shows the segmentation of all the items of this company. The thresholds used are arbitrary: 0.5 for CV and 100000 kilos for AV. Low CVs are easy to forecast and high CVs are hard to predict or in many cases, unforecastable. The division in AV shows the high and low impact items for the company. It is easy to see lots of items in the "long-tail" with low impact—that is, slow-moving items due to the uncontrollable tendency of marketers to apply inefficient innovations.

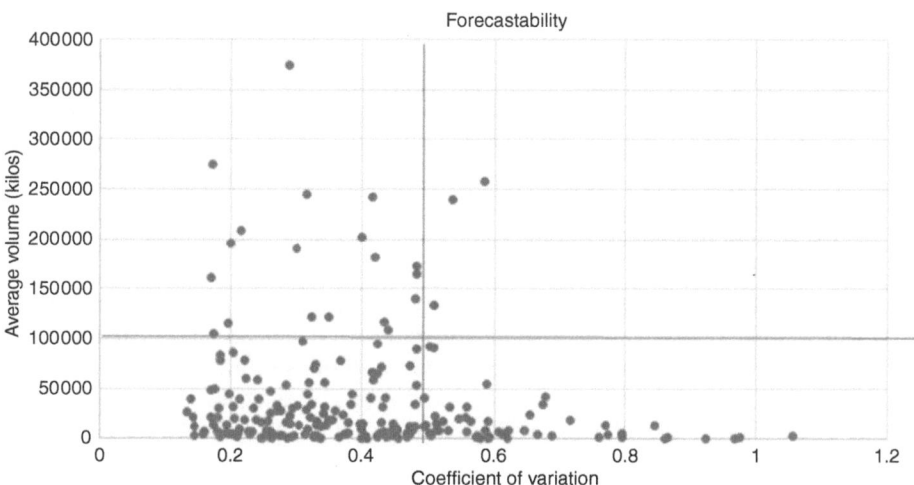

FIGURE 11.1 Forecastability for the products of a real company

The required procedure is concerned with predicting a dependent variable (sales) using a collection of independent variables (including the average price). The traditional technique to get this procedure is multiple regression, but with the inconvenience that it requires a model. Trends and seasonality can complicate it even further. The new alternatives are the Artificial Intelligence algorithms. The best results that have been reported in the literature is for the Random Forest Regressor (RFR). This was implemented and the results were quite acceptable. The code, the variables, and the data files required for it, are explained in Appendix 11A.

11.3.2 MODEL FOR PRESCRIBING DECISIONS

To support the decision making in an integrated supply chain, a linear mixed-binary mathematical model has been constructed as an approximation to the well-known non-linear model of a maximization process. The purpose is to use this model iteratively as an aid to the tactical and strategic decisions of the integrated S&OP (supply and demand) process. Among these decisions are:

- Which suppliers deal with the procurement of each material and in what quantities?
- Which plants produce the requirements of each distribution structure?
- What market share maximizes the profitability of the company in each zone segment, given the accomplishment of a market strategy represented by minimum participations?
- With what average prices (implemented by discounts and promotions) should each product be worked monthly in each market segment?
- What distribution structure will serve each segment?

All these decisions are made in monthly periods and for a horizon of one to more years. The mathematical model is presented in Appendix 11B. The logic of the model is composed of the following concepts:

- The sales of each market zone in each product are calculated based on an average price as well as the revenue obtained (linearization of the multiplication of the sales by the price), as will be explained later in the chapter.
- Average prices could be obtained from a demand elasticity formula or from the proposed procedure to get a relation between sales and prices. The average price values will be between a minimum and a maximum, determined according to market conditions.
- The marketing strategy is introduced by defining a minimum participation that is fulfilled outside the logic of maximization of the marginal contribution.
- Material balances are met for the three inventories of the model: materials in plants, finished products in plants, and finished products in distributors.
- The materials and the finished products warehouses of each plant, and the warehouses of each distributor, are restricted by storage capacities.
- The quantity to purchase per supplier for each material is restricted to a monthly maximum.

- The hours used by each production resource in each month are restricted by a capacity in available hours.
- The total amount to be distributed in each market zone is restricted by a capacity in units.

Figure 11.2 shows the base structure that represents the supply chain considered.

The objective of the model is to maximize the present value of the flow of monthly marginal contributions, given the accomplishment of a marketing strategy. The monthly marginal contribution is calculated by the difference between sales revenue and total costs—that is, purchases, transport, manufacturing, distribution, and inventory. The calculation of revenue from sales is the non-linearity that is present in every model that aims to maximize. The purpose is to have a practical model that can be used iteratively and thus, both the calculations of sales and of revenues based on price, are linearized. Economic theories demonstrate the existence of elastic and inelastic products. A product is elastic if the quantity demand of the product changes drastically when its price increases or decreases. Conversely, a product is inelastic if the quantity demand of the product changes very little when its price fluctuates. These relationships are fulfilled under perfect competition markets: that is, no buyer or seller can alter the price of a product. In real cases with randomness, any behavior can be present even in the same product, and for this reason a general linearization that works under any condition has been considered.

Figure 11.3 shows the sale and revenue variables based on the average price variable. If the company has developed demand elasticity formulas, a separable programming will be used that, although it is approximate, can become as accurate as the user requires by increasing the number of intervals used in the formulas. For each

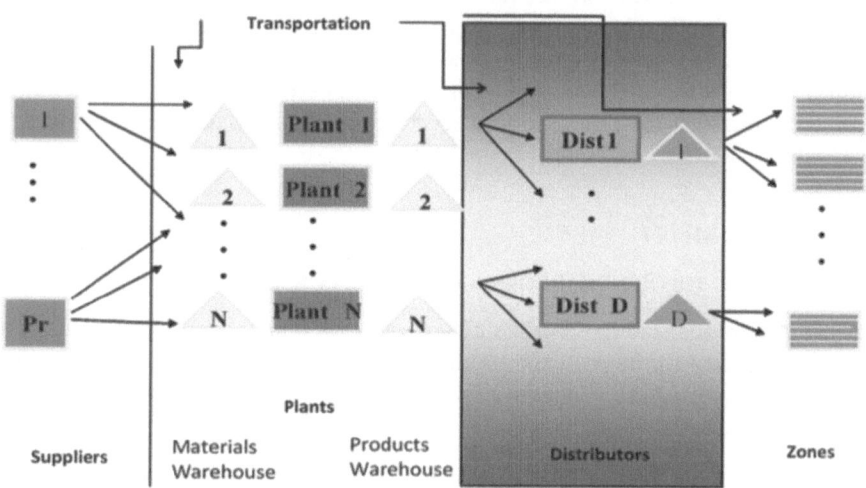

FIGURE 11.2 Base structure of the supply chain

Note: Pr: provisioners.

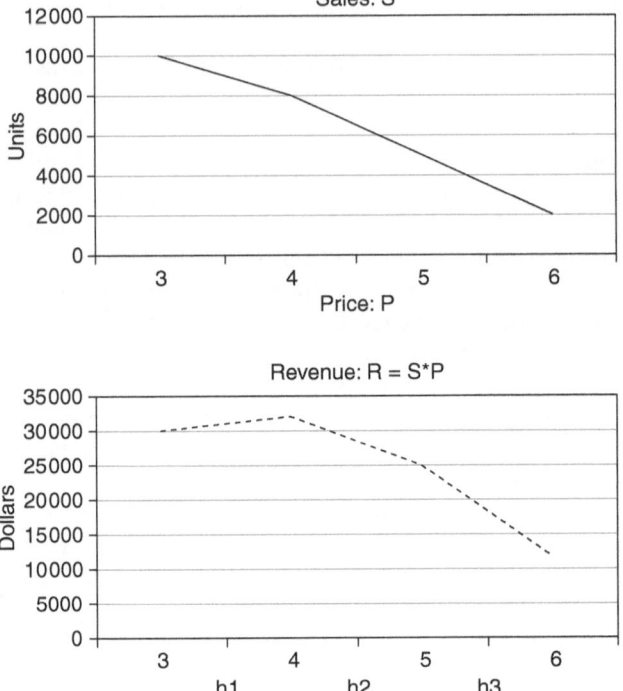

FIGURE 11.3 Linearization of sales and revenues based on average price

product in each zone and in each month, the two linear approximations shown are constructed: one for sales and one for revenue. As many intervals are used because precision is required, a binary variable appears for each of them that ensures the continuity of the values between intervals. In the author's more than 20 years of building solutions for SCM, only one company has used elasticity formulas. For all the other companies, the proposed procedure for predicting sales based on average prices should be used.

11.4 EXPERIMENT

11.4.1 Using the Random Forest Regressor in Real Data

The internal company data consists of six years of sale and billing records. The data was classified for items (each product in each zone). We chose three items to show the behavior of the procedure with different levels of CV but always able to be forecasted:

Item A (AV = 274867 kilos; CV = 0.1727)
Item B (AV = 47262 kilos; CV = 0.2618)
Item C (AV = 14675 kilos; CV = 0.2912)

TABLE 11.1
Results of RFR in Selected Items

Concept		Items A	B	C
Average monthly MAPE for 2018		4.71%	6.21%	8.13%
Coefficient of determination		0.7918	0.7942	0.5855
Importance	Year	0.04	0.18	0.11
	Month	0.15	0.21	0.12
	Sales_12	0.09	0.47	0.07
	Price	0.72	0.14	0.70

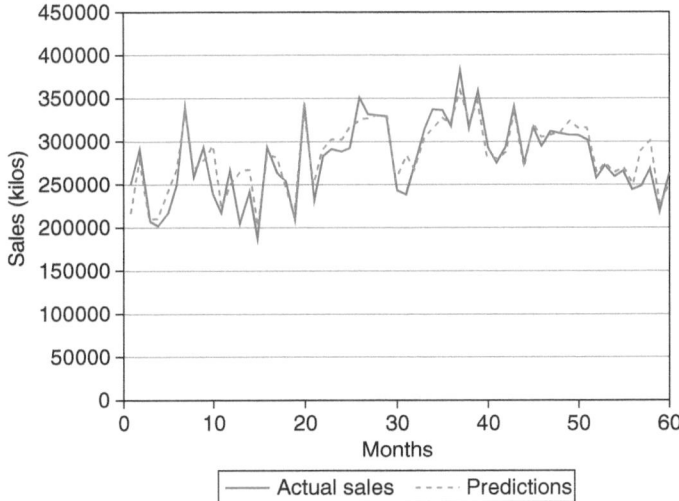

FIGURE 11.4 Actual sales and predictions for Item A

The forecasts obtained for these items will be used in the prescriptive model in Section 11.4.2. The performance metrics that the RFR obtained for these items are shown in Table 11.1. The mean absolute percentage error (MAPE) and the coefficient of determination for the three items are acceptable, but getting worse as the CV increases. The importance of the four independent variables used in the RFR (see Appendix 11A) shows behaviors quite different for the three items. Items A and C have an importance of more than 0.7 for the average monthly price. It demonstrates that the inclusion of this variable is not only to track the relation between sales and price, but also to achieve better forecasts. Item B shows a seasonal behavior, and it seems that the average price is not that important under this condition.

Figure 11.4 shows the plot of the actual sales and the predictions of the RFR for Item A. We can use this plot to determine if there are any outliers in either the data

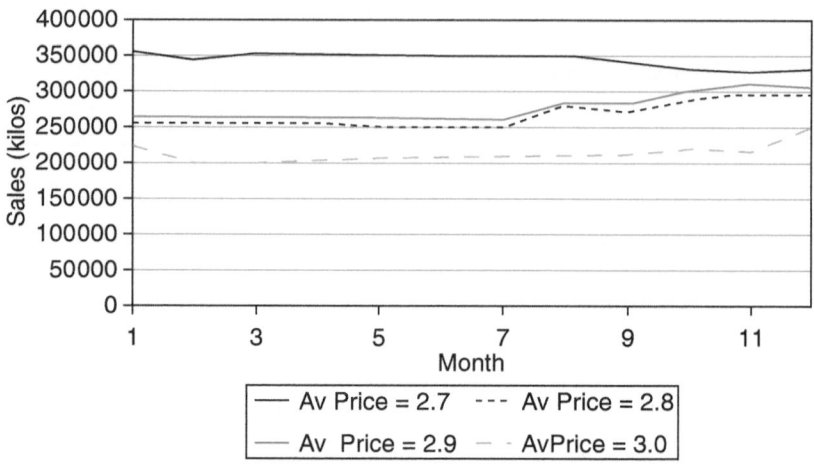

FIGURE 11.5   Sales forecasts for different average prices for Item A

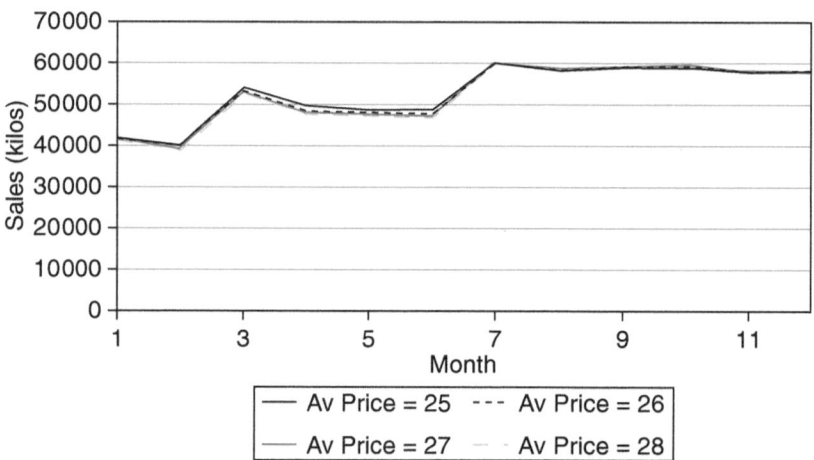

FIGURE 11.6   Sales forecasts for different average prices for Item B

or the predictions. The plot indicates no outliers and no bias. The model given by the RFR was used to forecast the next 12 months for values of the average price between the minimum and maximum that interests in the market. The results are in Figure 11.5 and they will be used in the prescriptive model. It shows a negative relation between sales and price but with different effect in different months. The levels of Av Price 2.8 and 2.9 are in contradiction to what could be expected, but this is something that can happen when you work with random real data.

With Item B the behavior is inelasticity—that is, the sales are not much affected by the average price, and this can be seen in Figure 11.6. Whereas Item C is a clear negative elasticity in two levels, as shown in Figure 11.7.

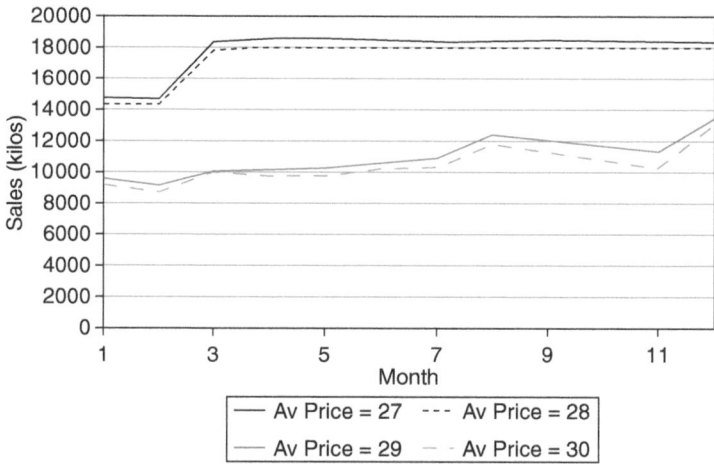

FIGURE 11.7   Sales forecasts for different average prices for Item C

11.4.2   Using Real Data in a Reduced Supply Chain

The mathematical model developed (see Appendix 11B) has been tested in multiple options with different quantities of products, market areas, suppliers, and average price alternatives for the corresponding items. A reduced version of two products (PR1 and PR2), two flexible production lines located in different plants, three distributors, and three market zones, was prepared to show the behavior of the proposed system.

The PR2 is much more profitable than PR1 in this version but with much less average volume per month. The three items considered in the experiment are used in the following way: Item A is the PR1 in each of the three zones Z1, Z2, and Z3; Item B is the PR2 in Z1 and Z2; and Item C is the PR2 in Z3. Running with these potential sales considered in the experiment and with a minimum participation of 70% for PR1 in all the zones, the following results are obtained.

The maximum present value of the monthly marginal contributions has the cash flow shown in Figure 11.8. No higher contributions are achieved because we limit the production capacity to show the desired behavior of the model. The model seeks the best combination of average prices and sales within this limit, maximizing marginal contributions but meeting the minimum sales stipulated for PR1. Figure 11.9 shows the sales and maximum demands of both products in market zone Z3. Figure 11.10 shows the proposed average prices for both products in the same zone Z3. In the two other zones, the proposed average prices for PR1 are the same and very close to the maximum price. For PR2 the proposed average price is irrelevant due to its inelastic demand. Since the objective function is to maximize the marginal contribution, the preferred product is PR2, but the model must meet the 70% of sales for PR1. In the first six months, Z3 is proposed to work with promotions for PR2 to obtain more margin contribution and in the rest of the months, except December, at the maximum price, because the limited production capacity must be used for PR1 and PR2 in Z1

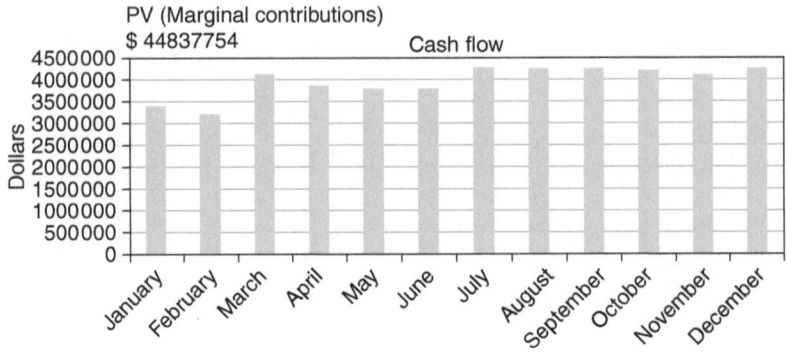

FIGURE 11.8   Flow with maximum marginal contribution

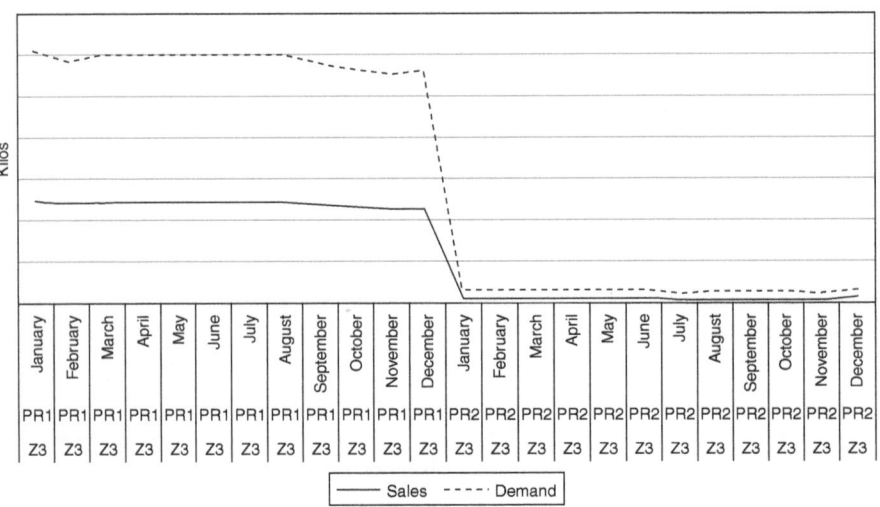

FIGURE 11.9   Sales and demand for PR1 and PR2 in zone Z3

and Z2. Remember that in Z1 and Z2 the average price for PR2 is irrelevant but the sales are much higher than in Z3, especially in the second semester. It can be seen how sales are accommodating between the two products to find the maximum contribution with the best average prices.

Relaxing the minimum market participation of PR1 to 60%, the marginal contribution increases to about $600,000 because the system can produce more of PR2, and therefore, the proposed average price for this product changes to the minimum level ($28) in the Z3 in all the months. For PR1 the participation is more than the required 60% in several months, with proposed average prices varying between $2.90 and $3.00.

These examples try to show the logic of this model in a reduced environment. With all the products of the company, the behavior is very similar to the example because

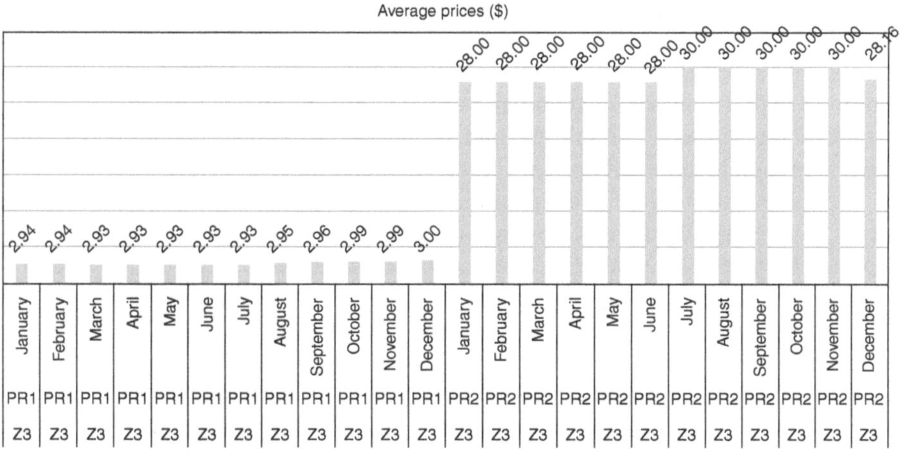

Average prices ($)

FIGURE 11.10 Average prices for products PR1 and PR2 in zone Z3

the competition for limited resources usually involves only groups of products. After these decisions, the supply managers have to fulfill the purchases, productions, and distributions committed; and the demand managers have to prepare a promotion plan to accomplish the average prices recommended.

11.5 CONCLUSIONS

This chapter presents a ML technique to forecast sales affected by average prices for B2C products, and a mathematical model to maximize the marginal contribution by simultaneously integrating the management of the supply side with that of the demand side, including pricing and marketing strategy, to support the S&OP process. Both supply (capacities) and demand are manageable in this proposal.

Currently, the disagreements between the areas of operations and marketing characterize the day-to-day business operation. The launch of promotions on products in which capacity is limited, the existence of inventories with low turnover, the presence of undesirable positive bias to protect sales, the existence of unused manufacturing resources due to lack of demand, inefficiencies and high costs due to demand variability, and the lack of profitability in products in which there is not sufficient capacity to satisfy demand, are all symptoms that demonstrate these disagreements. The proposed system seeks to eliminate these situations by integrating pricing and marketing actions of the demand side with the complex and changing conditions of the supply chain.

The use of the developed support will allow a company to prioritize its supply by profitability of the product or customer, given the accomplishment of a marketing strategy. The supply can be segmented into categories based on distribution and market channels. If the demand exceeds capacity, the logic of the model will determine the optimal mix of average prices and market shares to meet that demand with maximum profitability. When the demand does not cover the capacity, the same technique will make it possible to suggest decisions to stimulate demand (promotions) to reach a more efficient relationship with the supply. The adjustment will be permanent, so that average prices, market shares, and capacity allocations ensure the highest possible return on your assets. The system has been designed to run every month supporting the S&OP process with a simple and understandable procedure. The computing times are low enough to iteratively run any alternative that wants to be tested.

The achievement of the benefits of the system, although they can be proven from a quantitative aspect, could be somewhat difficult to accept from the organizational aspect. The sales and marketing areas are used to making autonomous decisions concerning pricing, sales quotas, promotions, market shares, and advertising, based on qualitative analysis, wrong incentives, and motivation policies. Pitfalls of this procedure have been noted. They can be avoided using systems like the one proposed, taking decisions in accordance with the results of the model and linking them to the S&OP process. In general, it is still necessary to look for appropriate ways in which strategic and tactical decisions recognize the proposed contributions. The research has shown that the well-known techniques of optimizing supply given a demand, or handling demand given a supply capacity, have a better destination if they are integrated and promotion decisions are incorporated in the presence of limited capacities, economies of scale, and seasonality of demand.

11.6   FUTURE RESEARCH

In the practical aspect, the developed system is ready to be incorporated with descriptive models that facilitate, through a decision support tool, its use as an aid to the S&OP process. The tool must include an environment for comparing tactical and strategic scenarios. To make strategic decisions, a procedure should be proposed in which "to be" scenarios must be compared with "as is" states. How to make these comparisons and how to support them is a subject of further investigation.

In the theoretical aspect, the consideration of risk in decisions is another research topic. All the decisions suggested in the present research have been taken based on the maximization of the present value of the company's profitability, given the accomplishment of a marketing strategy. It is logical to think that many companies may prefer lower returns if the associated risk is reduced. The way to measure these risks and how to alter the models so that they can find solutions other than those of maximum profitability using a return–risk relationship as their objective, remain unknown.

The proposed system is adequate for B2C companies where there exists a lot of promotions in product sales. Companies that sell products to other business (B2B) do not have enough variation in their pricing to calculate a price–sales relation. While pre-buying for volume discounts is possible, that does not affect the total volume bought over time. Another kind of solution is necessary for these companies.

REFERENCES

Bertsimas, D. and N. Kallus. 2020. "From predictive to prescriptive analytics." *Management Science*, 66 (3): 1025–1044.

Burrows, R. P. 2012. *The Market-Driven Supply Chain.* New York: American Management Association.

Cecere, L. M. and C. W. Chase. 2013. *Bricks Matter.* New Jersey: John Wiley & Sons.

E2open. 2019. "2019 Forecasting and Inventory Benchmark Study." White paper. www.e2open.com/the-2019-e2open-forecasting-and-inventory-benchmark-study/.

Ferreira, K., B. H. A. Lee, and D. Simchi-Levi. 2016. "Analytics for an online retailer: Demand forecasting and price optimization." *Manufacturing & Service Operations Management,* 18 (1): 69–88.

Fleishmann, M., J. Hall., and D. Pyke. 2011. "A dynamic pricing model for coordinated sales and operations." *SSRN Electronic Journal.*

Kaplan, U., M. Turkay, B. Karasozen, et al. 2011. "Optimization of supply chain systems with price elasticity of demand." *Informs Journal on Computing,* 23 (4): 557–568.

Lofvers, M. 2020. "SCM Survey: Software for end-to-end visibility in great demand." www.supplychainmovement.com/scm-survey-software-for-end-to-end-visibility-in-great-demand/.

Logility, Inc. 2018. "Three Checklists to Build a Successful Supply Chain Analytics Foundation." White Paper. www.sdcexec.com/sponsored-research/whitepaper/20999588/logility-three-checklists-to-build-a-successful-supply-chain-analytics-foundation.

Meissner, J. and A. Strauss. 2009. "Pricing Structure Optimization in Mixed Restricted/Unrestricted Fare Environments." Lancaster University Management School. https://eprints.lancs.ac.uk/id/eprint/48986/?template=browse.

Mišić, V. and G. Perakis. 2020. "Data analytics in operations management: A review." *Manufacturing & Service Operations Management,* 22 (1): 158–169.

Moon, M. A. 2013. *Demand and Supply Integration.* New Jersey: Pearson Education.

Padilla, J. A. 2014. "La evolucion del planeamiento de ventas y operaciones." *Interfases,* 7: 105–116.

Pedroso, C. B., L. D. R. Calache, F. R. Lima Junior, et al. 2017. "Proposal of a model for sales and operations planning (S&OP) maturity evaluation." *Production, 27.*

Scavarda, L., B. Hellingrath, T. Kreuter, et al. 2017. "A case method for sales and operations planning: A learning experience from Germany." *Production,* 27.

Senthilnathan, Samithamby. 2016. "The impact of elasticity on the firm's revenue." *International Journal of Science and Research,* 5 (9): 1728–1731.

Shapiro, Jeremy F. 2010a. "Advanced analytics for sales & operating planning." *Analytics*: 20–26.

Shapiro, Jeremy F. 2010b. *Beyond Supply Chain Optimization to Enterprise Optimization.* Cambridge, MA: Academic Research MIT.

Simchi-Levi, David. 2008. *Designing and Managing the Supply Chain.* New York: McGraw-Hill/Irwin.

Simchi-Levi, David. 2010. *Operations Rules: Delivering customer value through flexible operations.* Cambridge, MA: MIT Press.

Taskin Z., S. Agrali, A. Unal, et al. 2015. "Mathematical programming-based sales and operations planning at Vestel Electronics." *Interfaces,* 45 (4): : 325–340.

APPENDIX 11A: RANDOM FOREST REGRESSOR FOR THE REQUIRED FORECASTS

The following Python code for the Random Forest Regressor was used in the research.

```
1.  import pandas as pd
2.  read_file = pd.read_excel (r'path/salesdata.xlsx')
3.  read_file.to_csv (r'path/salesdata.csv', index=None, header=True)
4.  features = pd.read_csv ('path/salesdata.csv')
5.  print ('The shape of the data is:', features.shape)
6.  import numpy as np
7.  labels = np.array (features['Sales'])
8.  features = features.drop ('Sales', axis = 1)
9.  feature_list = list (features.columns)
10. features = np.array (features)
11. from sklearn.model_selection import train_test_split
12. train_features, test_features, train_labels, test_labels = train_test_split
    (features, labels, test_size = 0.2258, random_state = 42)
13. baseline_preds = test_features[:,feature_list.index ('Sales_12')]
14. baseline_errors = abs (baseline_preds – test_labels)
15. from sklearn.ensemble import RandomForestRegressor
16. rf = RandomForestRegressor (n_estimators = 1000, random_state = 42)
17. rf.fit (train_features, train_labels)
18. predictions = rf.predict (test_features)
19. errors = abs (predictions – test_labels)
20. print ('Mean Absolute Error:', round (np.mean (errors), 2), 'kilos')
21. mape = 100 * (errors / test_labels)
22. precision = 100 – np.mean (mape)
23. print ('Precision:', round (precision, 2), '%')
24. importances = list (rf.feature_importances_)
25. feature_importances = [(feature, round (importance, 2)) for feature, import-
    ance in zip (feature_list, importances)]
26. feature_importances = sorted (feature_importances, key = lambda x: x[1],
    reverse = True)
27. print ('The importances of the variables are:')
28. [print ('Variable: {:20} Importance: {}'.format (*pair)) for pair in feature_
    importances]
29. import datetime
30. year = features [:, feature_list.index('Year')]
31. month = features [:, feature_list.index('Month')]
32. dates = [str (int(Year)) + '-' + str (int(Month)) for Year, Month in zip (year,
    month)]
33. dates = [datetime.datetime.strptime (date, '%Y-%m') for date in dates]
34. true_data = pd.DataFrame (data = {'date': dates, 'Sales': labels})
35. year = test_features [:, feature_list.index ('Year')]
36. month = test_features [:, feature_list.index ('Month')]
```

37. test_dates = [str (int (Year)) + '-' + str (int (Month)) for Year, Month in zip (year, month)]

38. test_dates = [datetime.datetime.strptime (date, '%Y-%m') for date in test_dates]

39. predictions_data = pd.DataFrame (data = {'date': test_dates, 'prediction': predictions})

40. print ('The predictions for the test data are:')

41. print (predictions_data)

42. year = train_features [:, feature_list.index ('Year')]

43. month = train_features [:, feature_list.index ('Month')]

44. train_dates = [str (int (Year)) + '-' + str (int (Month)) for Year, Month in zip (year, month)]

45. train_dates = [datetime.datetime.strptime (date, '%Y-%m') for date in train_dates]

46. predictions = rf.predict (train_features)

47. predictions_data = pd.DataFrame (data = {'date': train_dates, 'prediction': predictions})

48. print ('The predictions for the train data are:')

49. print (predictions_data)

50. read_file = pd.read_excel (r'path/salesfore.xlsx')

51. read_file.to_csv (r'path/salesfore.csv', index = None, header = True)

52. features = pd.read_csv ('path/salesfore.csv')

53. print ('The shape of the forecast is:', features.shape)

54. feature_list = list (features.columns)

55. features = np.array (features)

56. year = features [:, feature_list.index ('Year')]

57. month = features [:, feature_list.index ('Month')]

58. future_dates = [str (int (Year)) + '-' + str (int (Month)) for Year, Month in zip (year, month)]

59. future_dates = [datetime.datetime.strptime (date, '%Y-%m') for date in future_dates]

60. predictions = rf.predict (features)

61. predictions_data = pd.DataFrame (data = {'date': future_dates, 'prediction': predictions})

62. print ('The predictions for the future sales are saved in the file ForeResult.txt')

63. with open ('path/ForeResult.txt', mode = 'w') as file_object

64. print (predictions_data, file = file_object)

To run this code, it is necessary to use two excel files:

- **SalesData.xlsx** with the data of the monthly sales and the variables to consider for the forecast.
- **SalesFore.xlsx** with the variables for the future months to make the forecasts considering the average prices alternatives.

Lines 1–5 acquire the data from the excel files and translate it to an accessible format. Lines 6–10 separate the data into the features and targets. The target, also known as the label, is the value we want to predict: in this case, the actual sales and the features are all the columns the model uses to make a prediction. Lines 11–12 randomly split the data into training and testing sets. Lines 13–14 establish a baseline (a measure that should be beaten with the model). If the model cannot improve upon the baseline, a different model should be tried. We use the sales from 12 months prior as the baseline. Lines 15–18 train the model and make the test predictions. Lines 19–28 determine performance metrics and calculate variable importance used to improve the model. Lines 29–41 print the test predictions. Lines 42–49 print the train predictions. Lines 50–55 acquire the data for future forecasts and translate it to an accessible format. Lines 56–64 calculate the forecasts using the RFR model and print them in the file ForeResult.txt.

In this research, we considered four independent variables after several improvements of the model: year, month, sales_12 (sales 12 months prior), and average price. The choice of the variables had to be limited because the model requires forecasts for 12 months. With more variables it is possible to get better results for the next month's forecast. To improve the model, we used the baseline, the mean absolute error, the accuracy, and the importance of the variables. The results of the forecasts are saved in a text file and then used in the prescriptive model.

APPENDIX 11B: MIXED-INTEGER MODEL FOR THE S&OP SUPPORT

Indices:

$i = product$

$z,v = market\ zone$

$m = month$

$p = manufacturing\ plant$

$r = production\ resource$

$s = supplier$

$j = material$

$t = interval\ (for\ the\ linearization)$

Data:

$int = monthly\ interest\ rate$

$CPCH_{sjm} = average\ cost\ of\ material\ m\ in\ supplier\ j\ during\ month\ m$

$CPR_r = cost\ of\ a\ production\ hour\ at\ resource\ r$

$CDT_v = unit\ cost\ of\ distribution\ at\ distributor\ v$

$RT_{ir} = production\ rhythm\ of\ product\ i\ at\ resource\ r$

$CIM_j = monthly\ inventory\ cost\ of\ material\ j$

$CIPP_i = monthy\ inventory\ cost\ of\ product\ i\ at\ any\ plant$

Data:

$CIDP_i$ = monthly inventory cost of product i at any distributor

CTM_{sjp} = transportation unit cost of material j from supplier s to plant p

CTP_{piv} = transportation unit cost of product i from plant p to distributor v

CTS_{viz} = transportation unit cost of product i from distributor v to market zone z

$PMIN_{zim}$ = minimal price of product i at market zone z during month m

$SMAX_{zim}$ = maximal sales of product i at market zone z during month m

h_{zimt} = price incremental for zone z, product i, month m between intervals $(t+1)$ and

SS_{zimt} = sales slope between intervals $(t+1)$ and t for zone z, product i, month m

RS_{zimt} = revenue slope between intervals $(t+1)$ and t for zone z, product i,
      month m

MSH_{zi} = minimum share for product i in market zone z

RQ_{ij} = requirement of materials j for each unit of product i

$RP(p)$ = set of production resources located in plant p

$MPCH_{sjm}$ = maximum supply of supplier s with material j during month m

MRS_{rm} = maximum production hours of resource r during month m

MDS_{vm} = maximum distribution units of distributor v during month m

FM_j = amount of material j in standard units of warehousing (palette)

FP_i = amount of product i in standard units of warehousing (palette)

MWM_{pm} = maximum storage of materials in standard units in plant p at
      month m

MWP_{pm} = maximum storage of products in standard units in plant p at month m

MWD_{vm} = maximum storage of products in standard units in distributor v
      at month m

Variables:

FL_m = contribution margin flow during month m (free variable)

AP_{zim} = average price of product i in zone z during month m

SL_{zim} = sales of product i in zone z during month m

RV_{zim} = revenue of product i in zone z during month m

IM_{pjm} = inventory of material j in plant p at the end of month m

IPP_{pim} = inventory of product i in plant p al the end of month m

IDP_{vim} = inventory of product i in distributor v at the end of month m

TM_{sjpm} = transported quantity of material j from supplier s to plant p during
      month m

TP_{pivm} = transported quantity of product i from plant p to distributor v during
      month m

TS_{vizm} = transported quantity of product i from distributor v to zone m during
      month m

Variables:

PCH_{sjm} = *purchased quantity of material j from supplier s during month m*
PR_{irm} = *production of product i on resource r at month m*
DT_{vim} = *distributed quantity of product i in distributor v during month m*
y_{zimt} = *auxiliary variable for each zone z, product i, month m and interval t*
B_{zimt} = *binary variable for each zone z, product i, month m and interval t*

All the variables are non-negative except the FL_m.

Model:

$$Max \sum_m FL_m / \left(1 + int^{(m-1)}\right)$$

$$FL_m = \sum_{z,i} \left(RV_{zim} - CDT_v * DT_{vim} - CIDP_i * IDP_{vim}\right)$$

$$- \sum_{s,j} \left(CPCH_{sjm} * PCH_{sjm} + CIM_j * IM_{pjm}\right)$$

$$- \sum_{r,i} CPR_r * \frac{PR_{irm}}{RT_{ir}} - \sum_{s,j,p} CTM_{sjp} * IM_{sjpm} - \sum_{p,i,v} CTP_{piv} * TP_{pivm}$$

$$- \sum_{v,i,z} CTS_{viz} * TS_{vizm} - \sum_{p,i} CIPP_i * IPP_{pim} - \sum_{v,i} CIDP_i * IDP_{vim} ; \forall m$$

Subject to,

$$AP_{zim} = PMIN_{zim} + \sum_t y_{zimt} ; \ \forall z,i,m \tag{1}$$

$$SL_{zim} = SMAX_{zim} + \sum_t SS_{zimt} * y_{zimt} ; \ \forall z,i,m \tag{2}$$

$$RV_{zim} = PMIN_{zim} * SMAX_{zim} + \sum_t RS_{zimt} * y_{zimt} ; \ \forall z,i,m \tag{3}$$

$$y_{zim1} \leq h_{zim1} ; \ \forall z,i,m \tag{4}$$

$$y_{zimt} \geq h_{zimt} * B_{zimt} ; \forall z,i,m,t \leq T-1 \tag{5}$$

$$y_{zim(t+1)} \leq h_{zim(t+1)} * B_{zimt} ; \ \forall z,i,m,t \leq T-1 \tag{6}$$

$$SL_{zim} \geq MSH_{zi} * SMAX_{zim} ; \ \forall z,i,m \tag{7}$$

$$IM_{pj(m-1)} + \sum_s TM_{sjpm} = IM_{pjm} + \sum_{i,(r \in RP(p))} RQ_{ij} * PR_{irm}; \quad \forall p,j,m \qquad (8)$$

$$IPP_{pi(m-1)} + \sum_{r \in RP(p)} PR_{irm} = IPP_{pim} + \sum_v TP_{pivm}; \quad \forall p,i,m \qquad (9)$$

$$IDP_{vi(m-1)} + \sum_p TP_{pivm} = IDP_{vim} + DT_{vim}; \quad \forall v,i,m \qquad (10)$$

$$DT_{vim} = \sum_z TS_{vizm}; \quad \forall v,i,m \qquad (11)$$

$$SL_{zim} = \sum_v TS_{vizm}; \quad \forall z,i,m \qquad (12)$$

$$PCH_{sjm} = \sum_p TM_{sjpm}; \quad \forall s,j,m \qquad (13)$$

$$PCH_{sjm} \le MPCH_{sjm}; \quad \forall s,j,m \qquad (14)$$

$$\sum_{iRT_{ir}>0} \frac{PR_{irm}}{RT_{ir}} \le MRS_{rm}; \quad \forall r,m \qquad (15)$$

$$\sum_i DT_{vim} \le MDS_{vm}; \quad \forall v,m \qquad (16)$$

$$\sum_j \frac{IM_{pjm}}{FM_j} \le MWM_{pm}; \quad \forall p,m \qquad (17)$$

$$\sum_i IPP_{pim} / FP_i \le MWP_{pm}; \quad \forall p,m \qquad (18)$$

$$\sum_i IDP_{vim} / FP_i \le MWD_{vm}; \quad \forall v,m \qquad (19)$$

The objective of the model is to maximize the present value of the flow of monthly contribution margins, given the accomplishment of a market strategy. Each monthly flow is calculated by the difference between sales revenue and total costs—that is, purchases, transport, manufacturing, distribution, and inventory. The calculation of revenue from sales is the non-linearity that is present in every model that aims to maximize. The sales and the revenue based on average price are linearized. Constraint (1) calculates the average price; constraint (2) calculates the sales; and constraint (3) calculates the revenue for each product in each market zone at each month. Constraints (4), (5), and (6) express the logic of a separable programming for the linearization with the help of the continuous variable y and the binary variable B. Constraint (7) allows a marketing strategy that should be accomplished by minimum participations for sales that are fulfilled outside the logic of contribution margin maximization. Material balances are met for the three locations for inventories of the model: materials in plants by constraint (8); finished products in plants by constraint (9); and finished products in distributors by constraint (10). Constraints (11), (12), and (13) are the balances of the transported quantities with the distribution, the sales, and the purchase processes, respectively. The maximum limits of the purchase, production, and distribution are controlled by constraints (14), (15), and (16) respectively. Finally, the warehousing capacities for the three inventories considered in the model are included in constraints (17), (18), and (19).

The model can be solved by any commercial solver.

12 Deep Neural Networks Applied in Autonomous Vehicle Software Architecture

Olmer Garcia-Bedoya, Janito Vaqueiro Ferreira, Nicolas Clavijo-Buritica, Edgar Gutierrez-Franco, and Larry Lowe

12.1 INTRODUCTION

It is estimated that around 1.2 million people die on roads in the world every year, and as many as 50 million are injured. Over 90% of these deaths occur in low- and middle-income countries (Garcia et al. 2018). The trend toward the use of automated, semi-autonomous, and autonomous systems to assist drivers has received momentum from major technological advances. But recent studies of accident rates have highlighted the challenges posed by autonomous and semi-autonomous navigation, and this has motivated researchers to take on studies in this field. Some critical issues when designing autonomous vehicles (AVs) are safety and security (Park et al. 2010). Nowadays, all levels of autonomy in vehicles use the application of Machine Learning (ML) (NHTSA 2013), and some examples are presented in Table 12.1.

This chapter aims to contribute to understanding how Deep Learning algorithms integrate into the AV architecture, identifying how these tasks are distributed in different hardware to integrate the reliability of automotive embedded systems with the flexibility of Linux architecture.

In the next section, the authors expand on the architecture of hardware and software of AVs to identify how distributed computing works, and where simulation helps to accelerate the generation of Deep Learning models. The following section presents a case study of Machine Learning in AVs to illustrate the concepts. Finally, conclusions are presented.

12.2 MATERIALS AND METHODS

This section discusses hybrid architecture where lightweight, real-time tasks, and computing-intensive tasks run on different, interconnected hardware. This research is part of the AV project VILMA01 (First Intelligent Vehicle of the Autonomous

DOI: 10.1201/9781003137993-13

TABLE 12.1
Examples of Machine Learning in Different Levels of Autonomy in Vehicles

Level of Autonomy	Example of Machine Learning Application
No automation (0)	The control of the vehicle is performed by the driver and Machine Learning helps in perceptions tasks to alert the driver.
Driver assistance (1) or partial automation (2)	Advanced driver-assistance systems (such as adaptive cruise control to control speed, and lane keep assist to aid steering) help with driving fatigue, although the driver must be ready to take control at any given moment. Machine Learning helps in perception tasks to jointly control systems and perform an assistant-driver function.
Conditional automation (3)	AVs are capable of driving themselves, but only under ideal conditions, and with limitations. The transfer is made through an emergency button or sophisticated algorithm like the google patent by Cullinane et al. 2014. Nowadays, it is a common approach in commercial vehicles, using Machine Learning both in perception and in other tasks of the vehicle. In this stage, the data re-collected is crucial to the development of Deep Learning algorithms.
Full automation (5)	AVs have the capability to navigate in multiple environments, making it necessary that automation can adapt to different conditions. This has made Deep Learning and other Machine Learning techniques crucial to the development of this type of technology.

Mobility Laboratory) (Garcia et al. 2018) developed at UNICAMP, Brazil. The purpose of this vehicle is to enable the testing of cooperation strategies between the driver and the automated system.

Taking account of the definition of functional groups of AVs (Garcia and Diaz 2018), the design and development of automation for a street-approved car (a Fiat Punto 1.4V *Attractive* vehicle) began. The automation consisted of controlled steering, accelerator, brake, and the gearbox of the vehicle by an embedded system. In the steering, the automation was carried out through an electric motor coupled to the steering axle without making mechanical modifications so as not to affect the approval of the vehicle. The other three systems were automated by electronic circuits and data networks. For the development of automation, it was defined that every programmable device should have a debugging and programming procedure. This procedure should include the required hardware and software elements. Next, we will discuss software architecture to understand where Deep Learning software can interact with the vehicle.

12.2.1 Software Architecture

The software architecture was designed so that the AV modules had the ability to be developed in three different programming architectures, as described below:

1. Firmware software developed directly on the hardware of electronic equipment. At the moment, the only one modified is the steering motor software, since the other firmware within the architecture is programmed by third parties but must be configured and therefore known.
2. Embedded software implemented in the embedded system designed for real-time tasks in charge of interacting with hardware and firmware. The programs designed in this architecture are state machines from the automation hardware iteration with the software, an emergency routine, control layer routines that have real-time requirements, location routine, and interaction with sensors. Note that there is also a program that allows communication with the software on Linux.
3. Software on Linux is implemented in PC-type architecture with the possibility of being distributed on different computers which also contemplate the simulation. The first requirement to design the architecture was to define the mediator (middleware) or meta-operating system where all the programs of the PC-type architecture will be executed.

Tools were studied to make the transfer of information between the system modules robust, distributed, easy to use, and capable of exchanging information synchronously and asynchronously, such as KogMo-RTDB (Geiger et al. 2012), Microsoft Robotics Developer Studio, and the Robotic Operating System (ROS) (Quigley et al. 2009). ROS was the mediator selected to make connections between the programs. Having defined ROS as *middleware* of the software, on Linux, the programs to connect the vehicle's existing instrumentation and the VILMA01 embedded system were implemented and tested. Multiple AV architectures adopt the ROS framework for the development of perception and navigation programs (Belcarz et al. 2018; Kato et al. 2015; Marin-Plaza et al. 2018; Bruno et al. 2018), which has allowed us to take advantage of tools and programs developed under open-source licenses. Figure 12.1 illustrates the software architecture of the VILMA01, where we show how each functional block interacts.

In Figure 12.1, the light shading is the firmware software, the dark shading is the embedded software, and no shading indicates distributed software which can be implemented in many computers. This chapter concentrates on the functional block of Artificial Vision which is implemented in a computer with a graphics processing unit (GPU), which uses high computational resources.

12.2.2 CONVOLUTIONAL NEURAL NETWORKS

A short illustrative example is now discussed to exemplify the derivation and implementation of convolutional neural networks (CNNs) (see also LeCun et al. 1998). A CNN involves more connections than weight. This neural network is designed to learn filters in a data-driven manner. The example is for two-dimensional data and convolutions, but it can be extended to a greater number of dimensions. This description begins with a short discussion of backpropagation (see, e.g., Bishop 1995 for more details). The error was made by:

$$E^N = \frac{1}{2} * \sum_{n=1}^{N} \sum_{k=1}^{c} \left(t_k^n - y_k^n \right)^2 \tag{1}$$

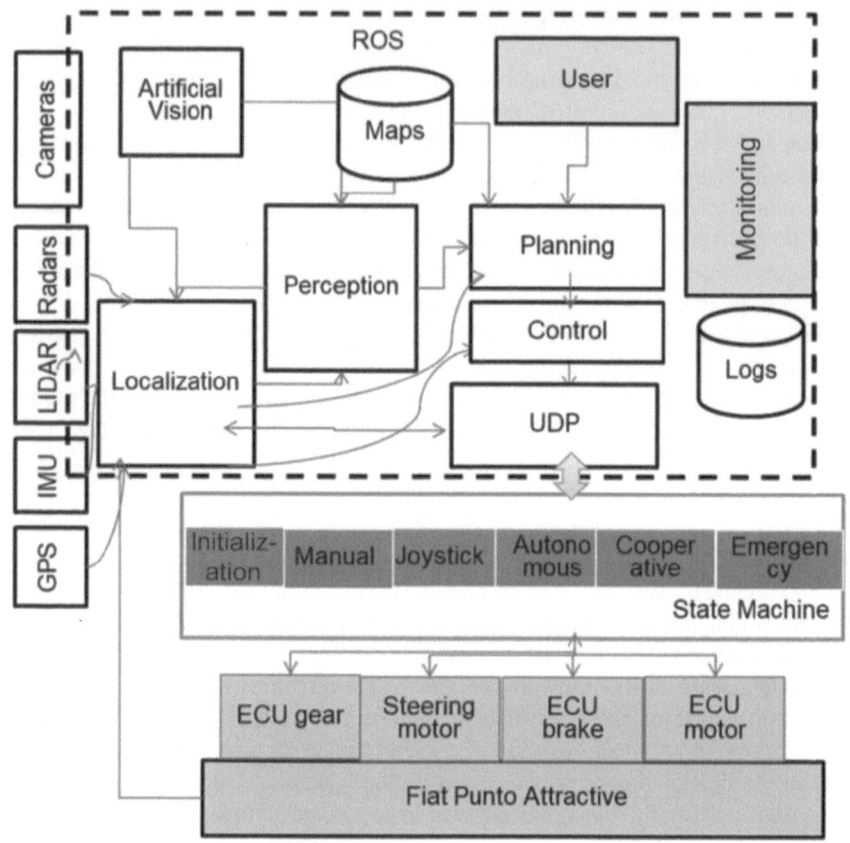

FIGURE 12.1   Software architecture of the blocks that comprise VILMA01

Where t_k is the k-th dimension of the n-th pattern's matching target (label), and y_k is the value of the k-th output layer unit in response to the n-th input pattern. The next formula denotes the current layer, with the output layer designated to be layer l+1 and the input "layer" designated to be layer 1. The activation function f () is usually selected to be the logistic (sigmoid) function f (x) = (1 + e−βx)−1 or the hyperbolic tangent function f (x) = a tanh(bx).

$$x^l = f(u^l), u^l = W^l x^{l-1} + b^l \qquad (2)$$

The "errors" which are propagated backward through the network can be thought of as "sensitivities" of each unit concerning perturbations of the bias:

$$\frac{\partial E}{\partial b} = \frac{\partial E}{\partial u}\frac{\partial u}{\partial b} = \delta \qquad (3)$$

with:

$$\delta^l = \left(W^{l+1}\right)^T \delta^{l+1} \tag{4}$$

$$\frac{\partial E}{\partial W^l} = x^{l-1}\left(\delta^l\right)^T \tag{5}$$

Convolutional layers are combined with sub-sampling layers (Serre et al. 2005). In general:

$$x_j^l = f\left(\sum_{i \in M_j} x_i^{l-1} * k_{ij}^l + b_j^l\right) \tag{6}$$

where Mj characterizes a selection of input maps. The process repeats the same calculation for each map j in the convolutional layer, pairing it with the corresponding map in the sub-sampling layer:

$$\delta_j^l = \beta_j^{l+1}(f'\left(u_j^l\right) \tag{7}$$

Now follows the calculation for the bias gradient by summing over all the entries in ∂j. Finally, the gradients for the kernel weights are computed using backpropagation. Following a sum over the gradients:

$$\frac{\partial E}{\partial b_j} = \sum_{u,v} \delta_{ju,v}^l \tag{8}$$

$$\frac{\partial E}{\partial k_{ij}^l} = \sum_{u,v} \delta_{ju,v}^l \left(p_i^{l-1}\right)_{u,v} \tag{9}$$

where $\left(p_i^{l-1}\right)_{u,v}$ represents the patch that was multiplied element-wise during convolution in order to compute the element at (u, v) in the output convolution map. A sub-sampling layer produces down-sampled versions of the input maps. With N input maps, there will be N output maps, and where down() represents a sub-sampling function.

$$x_j^l = f\left(\beta_j^l down\left(x_j^{l-1}\right) + b_j^l\right) \tag{10}$$

The learnable parameters to update are the bias parameters β and b.

$$\frac{\partial E}{\partial \beta_j} = \sum_{u,v} \left(\delta_j^l \right)_{uv}$$

(11)

The multiplicative bias β involves the original down-sampled map computed at the current layer during the feedforward pass (Bouvrie 2006).

$$d_j^l = down\left(x_j^{l-1} \right)$$

(12)

Then the gradient for β is given by:

$$\frac{\partial E}{\partial \beta_j} = \sum_{u,v} \left(\delta_j^l \circ d_j^l \right)_{uv.}$$

(13)

In order to set up the process to learning combinations, let αij be the weight given to input map i when forming output map j.

$$x_j^l = f\left(\sum_{i=1}^{N_{in}} \alpha_{ij} \left(x_i^{l-1} * k_i^l \right) + b_j^l \right)$$

(14)

where:

$$\sum_i \alpha_{ij} = 1, \quad and \quad 0 \le \alpha_{ij} \le 1.$$

The derivative of the softmax function is given by:

$$\frac{\partial \alpha_k}{\partial c_i} = \delta_{ki} \alpha_i - \alpha_i \alpha_k$$

(15)

$$\frac{\partial E}{\partial \alpha_i} = \frac{\partial E}{\partial u^l} \frac{\partial u^l}{\partial \alpha_i} = \sum_{u,v} \left(\delta^l \circ \left(x_i^{l-1} * k_i^l \right) \right)_{uv}$$

(16)

Finally, it is possible to use the chain rule to compute the gradients of the error function (1) with respect to the underlying weights.

12.2.3   CONVOLUTIONAL NEURAL NETWORKS EXAMPLE

When detecting the complexity of a scenario using CNNs, one of the most significant issues is how to deal with images and videos in simulation. But technology has been advancing at a fast pace: in recent years, Deep Learning models (specifically CNNs) have been able to provide potential solutions. By utilizing a CNN Deep Learning engine, we can automatically analyze the identification of battle scenarios.

We know that when a computer sees an image, what it actually "sees" is an array of the numbers that identify the pixels. For example, an image can be 100 x 100 x 3 in size. The number 3 refers to RGB (color) values. For RGB, each of its dimensions (i.e., 3) has a value from 0 to 255. These values tell us about the pixel intensity at each point. These numbers are the only inputs the computer uses. Our goal is to leverage the computer to classify this image automatically. For our application, it is the probability that a certain image that describes the battle scene can be classified as low or high risk for the operator of one of the artifacts that are part of the simulation. In our case, this is from the perspective of the weapon operator. The satellite imagery is evaluated with respect to the margin of error. For a low-risk image, the pilot is expected to survive, while a high-risk image is expected to lead to failure.

CNNs have become very popular in the last five years (Elhassouny and Smarandache 2019; Yao, Wu, and Gao 2020; Passalis, et al. 2020): they are able to recognize images, and they are expected to be the trend in technology to provide breakthroughs in the short future. A CNN has several steps: Convolution, Pooling, Flattening, and Full Connection.

As part of our research, we decided to encapsulate the knowledge of scenarios using a CNN architecture. This endeavor resulted in a challenging engineering task. There are many parameters utilized in the development of a CNN, and these parameters must be unique to the image classification problem at hand. One small/medium-sized enterprise (SME) was selected to classify the situations from images from the battlefield. After numerous designs and trials, a CNN was built, which has seven layers and thousands of neurons (i.e., a very complex architecture). Figure 12.2 displays the architecture created for this problem.

12.2.3.1 Training Workflow

It was decided to use KNIME® to build this complex architecture. A very sophisticated workflow was produced that not only created the CNN but involved several image processing steps. First, we have to read the files of the different scenarios played in SIMbox, which is accomplished by the "List Files" node. These images depicted features of the scenes and were classified by an SME (see Figure 12.3).

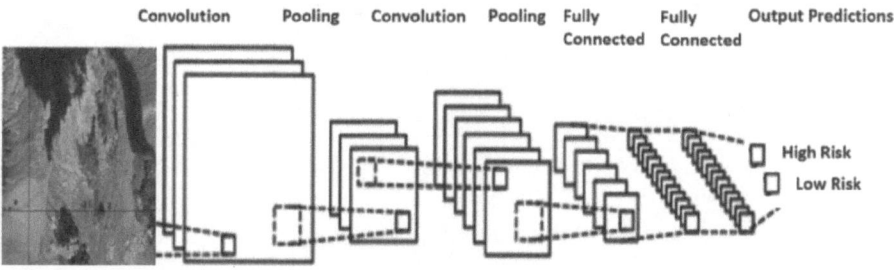

FIGURE 12.2   CNN developed to classify training scenarios as high and low risk

Then, to reduce the level of complexity, the three channels (RGB) were alleviated by getting an average of the numbers through the "Projector" node. The images were reduced to 150 x 150 x 1 in the "Image Resizer" node. Next, the pixels given in integers were converted to floating-point numbers to facilitate calculations in the "Image Converter" node. Now, the floating numbers were normalized between 0 and 1. The mean of the pixels of each image was calculated using the "Feature Calculator" node, and then subtracted from every pixel in the "Image Calculator" node. This is done to "center" the data. In the training process, you need to add the weights multiplied by the activations and then add the biases. And from the activations of the higher layers that later backpropagate with effective gradients. By entering this data, there is more homogeneity and the gradients are stable. Also, a single learning rate will be required. Now, each image is labeled with *high risk* or *low risk* through the "Rule Engine" node.

Figure 12.4 shows a detailed description of the CNN developed, which has seven layers. The first layer is the input layer that has 10,000 units. The second layer is a convolutional layer that uses 32 kernels, kernel size of 3,3, and a stride of 2,2 with ReLU (Rectified Linear Unit) units. The third layer is a sub-sampling layer that has a pooling function of a max kernel size of 2,2, and a stride of 2,2. The fourth layer is another convolution layer with 64 kernels, kernel size of 3,3, and a stride of 2,2. The fifth layer is again a sub-sampling layer that has a pooling function of a max kernel size of 2,2, and a stride of 2,2. The next two layers are dense: the sixth layer has activation ReLU, and the seventh layer has an activation sigmoid. Finally, there is an output layer with two neurons: one for low risk and the other one for high risk.

The optimization function is very sophisticated and very efficient for classification architecture. The optimization is stochastic gradient descent (SGD). The "updater" is RMSprop. The loss function is a negative log-likelihood.

SGD is an iterative method for optimizing an objective function with differentiable properties. It can be said that SGD is a stochastic approximation of gradient descent optimization. This relationship is noted since it replaces the actual gradient with an estimate of the gradient (calculated from a randomly selected subset). Especially in high dimension optimization problems, this reduces the computational load, achieving faster iterations for a slower convergence rate (Bousquet 2012).

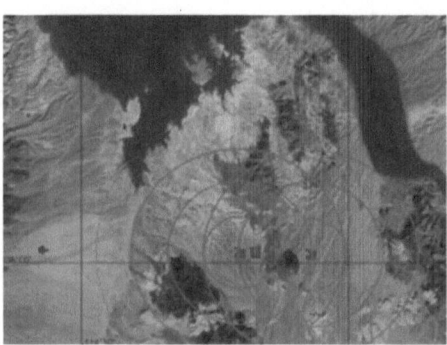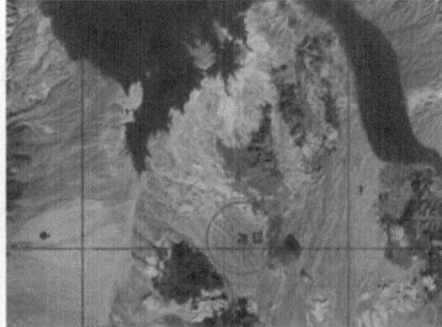

FIGURE 12.3 Examples of a high-risk scenario *(left)* and a low-risk scenario *(right)*

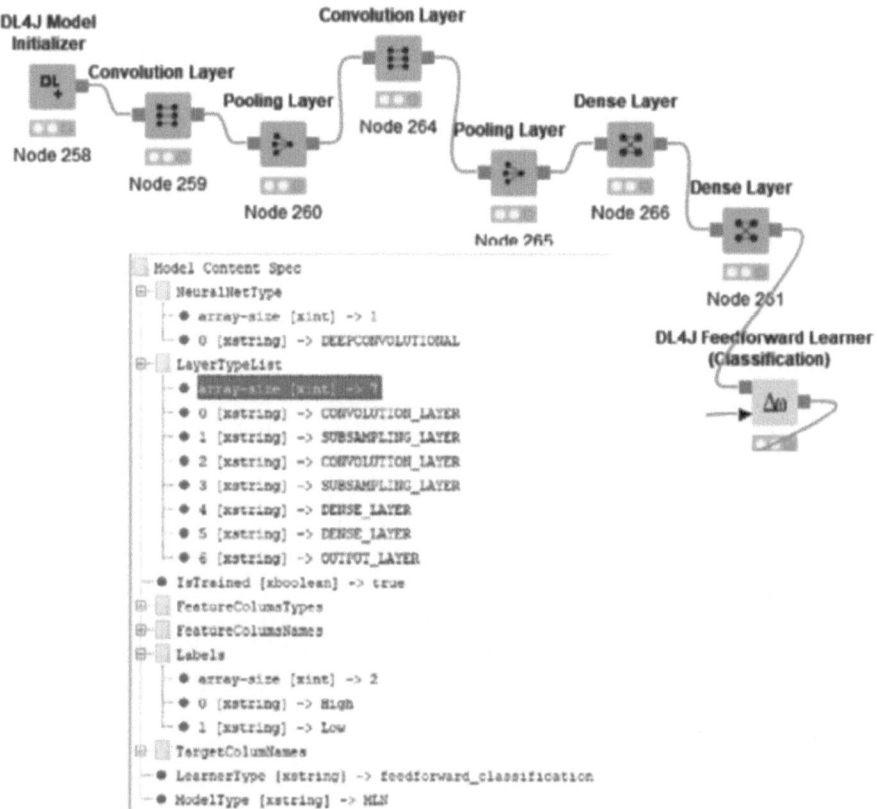

FIGURE 12.4 Building the CNN with seven layers and thousands of neurons

On the other hand, RMSprop (see Mosca and Magoulas 2015; Karpathy 2017; Wilson et al. 2017) is a mechanism to update the parameters (weights). The RMSprop is used to keep moving the average of the squared gradients for each weight. Consequently, the gradient is divided by the square root of the mean square, which is why it is called RMSprop (root mean square). The resulting rule is represented by the following formula (Bushaev 2018):

$$W_t = W_{t-1} - (\eta \left(\frac{\delta C}{\delta W} \right) / \sqrt{E(g^2)}) \qquad (17)$$

Where, W_t is the weight at time t, W_{t-1} is the weight at time t-1, η is the learning rate, $\delta C/\delta W$ is the derivative of the loss function by the derivative of the weight, and $E(g^2)$ is the moving average of the squared gradients. The negative log-likelihood is like minimizing the cross-entropy of a function.

In a CNN, the objective of the learning process (that is, to obtain weights) is nothing more than to minimize the difference between the distribution that we calculate and the actual distribution of the data. According to De Boer and colleagues (2005), the cross-entropy between two probability distributions over a set of observations measures the average number of information needed to identify an event if the encoding follows a learned probability distribution rather than the right but unidentified distribution. However, the actual probability distribution of the data is hidden and replaced by the empirical probability distribution over a training set with the characteristics of independence and withdrawal in an intrinsic uniformity (Vapnik 2000). As is very well stated by Zhu and colleagues (2018) and Goodfellow, Bengio, and Courville (2016), under the assumptions of classification, minimizing cross-entropy is equivalent to maximum probability—that is, "the learning problem aims to maximize the probability of a correct class for each of the training samples" (Zhu et al. 2018: 1). Maximum likelihood is a general training criterion by which the model learns the probability of the correct class for observation. Therefore, Zhu and colleagues (2018) state that the model makes predictions using Bayes' rules to calculate the posterior probabilities of the target classes for the observed data sample, and then selects the most likely classification label. These are the principles of negative log-likelihood.

The training of the network was performed by 10 epochs with several iterations and using mini-batches as a way to take advantage of the optimization process. Figure 12.5 displays the training process with the respective errors and convergence. The testing also shows excellent performance, and the CNN developed is capable of identifying the features of the battlefield from images, and classifying it, with a good level of accuracy.

The final node is the "Scorer," which reviews the predicted result and compares it to the actual category. This node produces two outputs: the "Confusion Matrix", and the accuracy measures. Table 12.2 shows that out of 14 possible entries that were categorized as high risk by our SME, 2 of them were incorrectly categorized as low risk. It also shows that out of 33 possible low-risk entries categorized by our SME, 3 of them were incorrectly

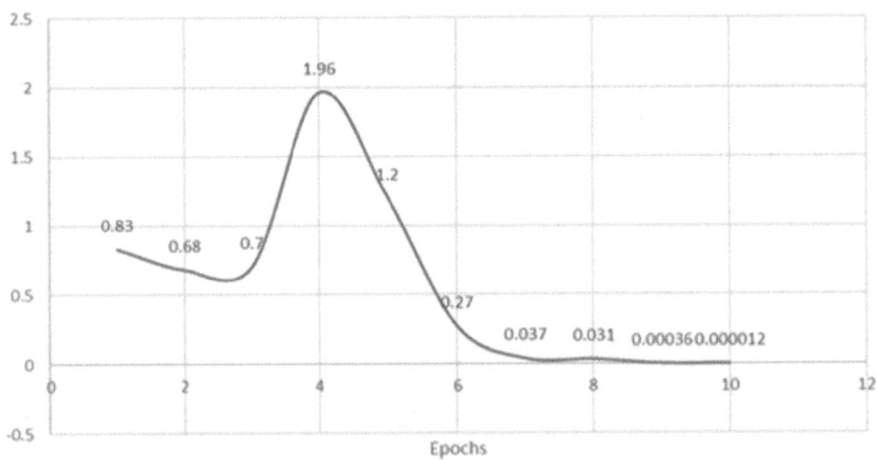

FIGURE 12.5   Convergence in 10 epochs for the CNN developed

TABLE 12.2
CNN Confusion Matrix

Row ID	High	Low
High	12	2
Low	3	30

categorized as high risk by the CNN. These numbers show that CNN is correctly categorizing a large majority of the cases, and this can be improved with more development. The errors can be reduced if additional images are processed. For the remaining measures displayed in Table 12.3, we are primarily interested in the "Accuracy" column. In this instance, the model ran with 89.4% accuracy.

12.3 RESULTS FOR AUTONOMOUS VEHICLES WITH DEEP NEURAL NETWORKS

Perception tasks in AV represent many challenges (Schwarting, Alonso-Mora, and Rus 2018); one of them being the requirement of big computational resources, which made a significant task to test our approach of integrating in the architecture. One of the key elements of perception is about the learning process to identify and classify traffic signals. The case is adapted from Hassan (2020) and uses the public dataset from the German Traffic Recognition Benchmark (GTSRB), which is a multi-class image classification benchmark in the domain of AV (Stallkamp et al. 2020). The experiments were conducted using Python 3.7® and Machine Learning packages. Additionally, the experiments were deployed according to parameters of corroboration, such as validation and test ratios of 20% each. In parallel, the value of epochs was 10, with a batch size of 50 for traffic signals images that must be validated. Afterwards, the CNN model was developed with 60 filters and 500 nodes in hidden layers. The implementation was carried out in four phases plus a final phase that is responsible for the implementation of the AV, considering its embedded system (Figure 12.6).

12.3.1 DATA ANALYSIS

The database consists of a set of images where each traffic signal ID represents a file that describes each signal. Both the image set and the labels are computationally represented as arrays. The first array contains features from the image database, such as traffic signal pixels. The second contains a type of traffic signal. Both contain a number of samples. An additional array represents a border surrounding the traffic signal image (see Figure 12.7). The most significant contribution of the data analysis phase is the recognition of the image database and how to process it. (Figure 12.7 shows some randomly selected samples).

Summarizing, the data analysis allows us to identify that this is a multi-class classification problem, with 43 classes of traffic signals to evaluate (based on Stallkamp

TABLE 12.3
CNN Accuracy Measures

Row ID	True Positives	False Positives	True Negatives	False Negatives	Recall	Precision	Sensitivity	Specificity	F-Measure	Accuracy	Cohen's Kappa
High	12	3	30	2	0.857	0.8	0.857	0.909	0.828		
Low	30	2	12	3	0.909	0.938	0.909	0.857	0.923		
Overall										0.894	0.751

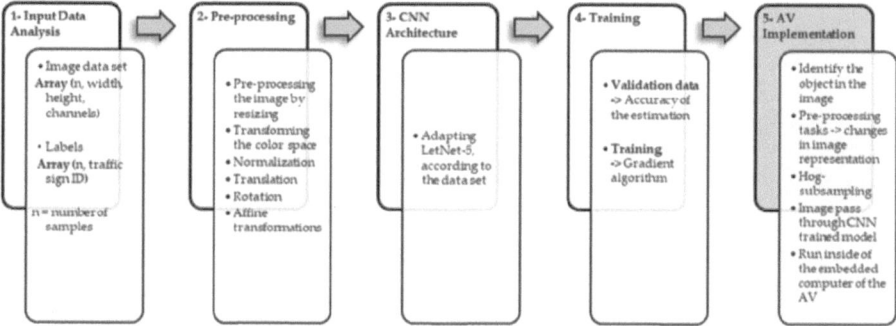

FIGURE 12.6 Phases of the CNN implementation of AVs

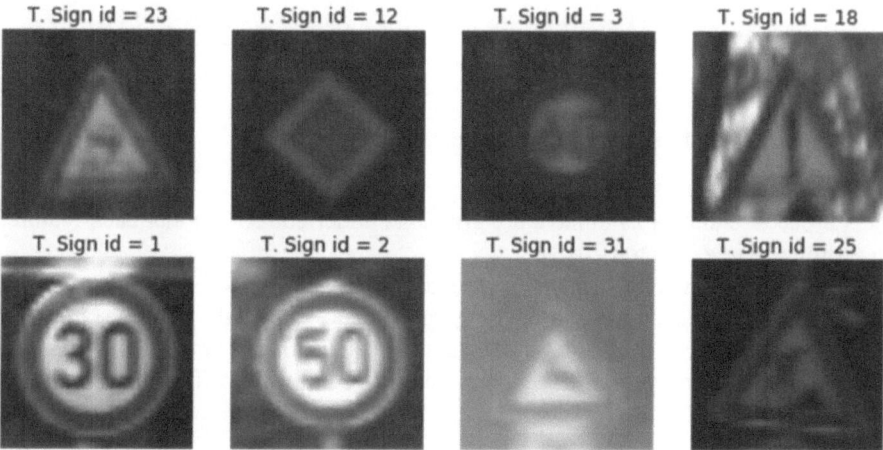

FIGURE 12.7 A random sample of the dataset

et al. 2020). Each image represents a single signal and can range in size from 15 x 15 to 250 x 250 pixels. Additionally, the disposition of each image does not always have the traffic signal in the center. Similarly, it may contain borders that span up to 10% of the total image size.

12.3.2 PRE-PROCESSING AND DATA AUGMENTATION

The pre-processing phase contemplates a series of tasks that can help to improve the subsequent process of training the network. Among the processing practices, we can highlight: (1) changing the image size (32 x 32 is common in the literature); (2) converting the color space by transforming to grayscale or RGB format; and (3) normalizing the data to scales that are compatible with the neural networks (ranges between 0 and 1 or between -1 and 1). An alternative to achieve normalization is to divide each dimension by its standard deviation, as long as it is zero-centered.

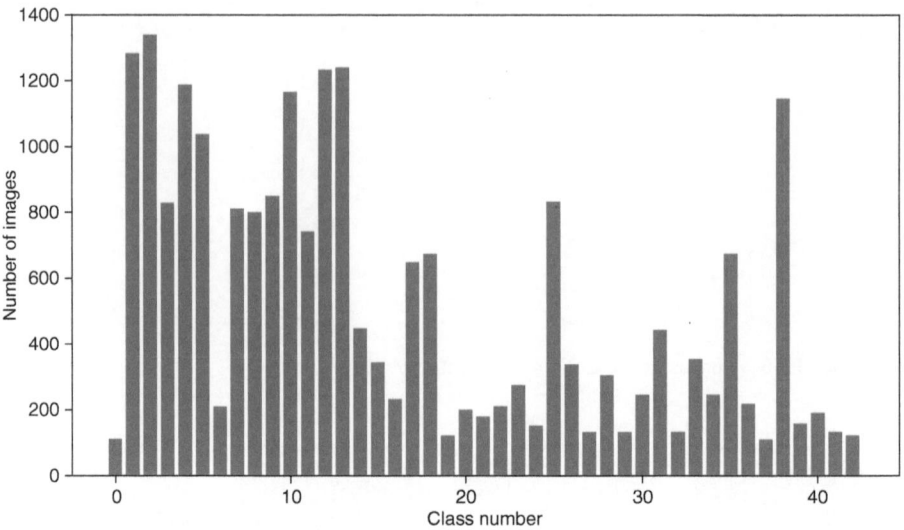

FIGURE 12.8   Distribution of the training data set

This procedure allows us to have the gradients under control, so that each feature is in a similar range (Heaton 2013).

Figure 12.8 illustrates that there are different numbers of each sample in the training data set—meaning heterogeneity in the data. This might generate overfitting. A common practice to deal with this is to generate new images, or to alter some images randomly through operations such as translation, rotation, and affine transformations.

12.3.3   CNN ARCHITECTURE

Based on the architecture of CNN proposed by LeCun and colleagues (1998), called LeNet-5, an adaptation is made to address the problem of recognition of traffic images in the context of their subsequent use in an architecture of operation in an AV (see Section 12.3.4). Figure 12.9 illustrates this adaptation.

The layers used in the adapted case are shown in Table 12.4. The layers and hyper-parameter values can be modified across the training process by using techniques from operations research, and heuristics and metaheuristics approaches, among others. The hyper-parameters such as K = number of filters, F = filter size (F x F), S = stride, and P = amount of padding, are relevant for CNN and can take different values for each layer through the iterative tuning process.

The training process is deployed by the following procedure. Split the training data between training and validation. Validation data is used for calculating both the accuracy and the loss of the estimation (see Figure 12.10).

In addition, the optimization model used in this algorithm is based on gradient descent stochastic optimization (Kingma and Ba 2015). Table 12.5 details the training algorithm (pseudocode).

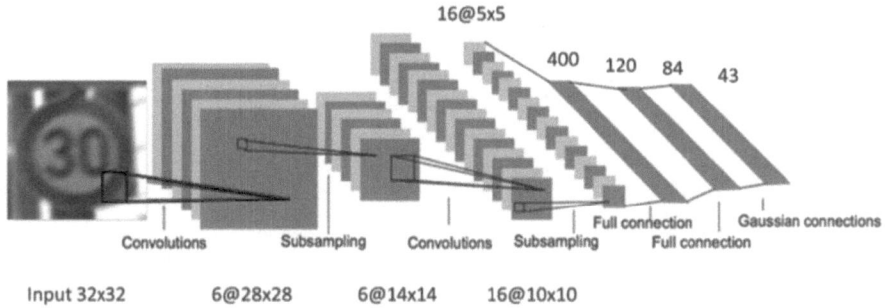

FIGURE 12.9   Adaptation of the architecture of a CNN (LeNet-5): each plane is a feature map, i.e., a set of units whose weights are constrained to be identical

Source: Adapted and modified from LeCun et al., 1998.

TABLE 12.4
CNN Layers (Adapted from LeNet-5)

Layer	Operation	Input	Output
1	Convolutional*	$32 \times 32 \times 1$	$28 \times 28 \times 6$
2	Sub-sampling max-pooling	$28 \times 28 \times 6$	$14 \times 14 \times 6$
3	Convolutional*	$14 \times 14 \times 6$	$10 \times 10 \times 16$
4	Sub-sampling max-pooling	$10 \times 10 \times 16$	$5 \times 5 \times 16$
5	Flat layer, 3-D to 1D	$5 \times 5 \times 16$	400
6	Fully connected layer*	400	120
7	Fully connected layer*	120	84
8	Output layer+	84	43

Note: *Activation of the ReLU function; +Activation of the Soft-max function.

12.3.4   AUTONOMOUS VEHICLE IMPLEMENTATION

The tasks of the Deep Learning pipeline use the cloud due to the requirements of high computational resources. The training algorithm needs to be integrated inside the architecture of the AV because the trained model is used to tag the images in real-time into testing scenarios. The first task in a real-time scenario is to segment the image to identify where the algorithm will look for the traffic signal. Consequently, the next pipelines are performed:

1. The image is received like a topic inside the node which identifies the object in the images.
2. Some pre-processing tasks performed in Opencv® are made, such as histogram equalization. It allows changes in the image representation from RGB to greyscale, and in terms of HSV (hue and saturation values). Even filters would be included.

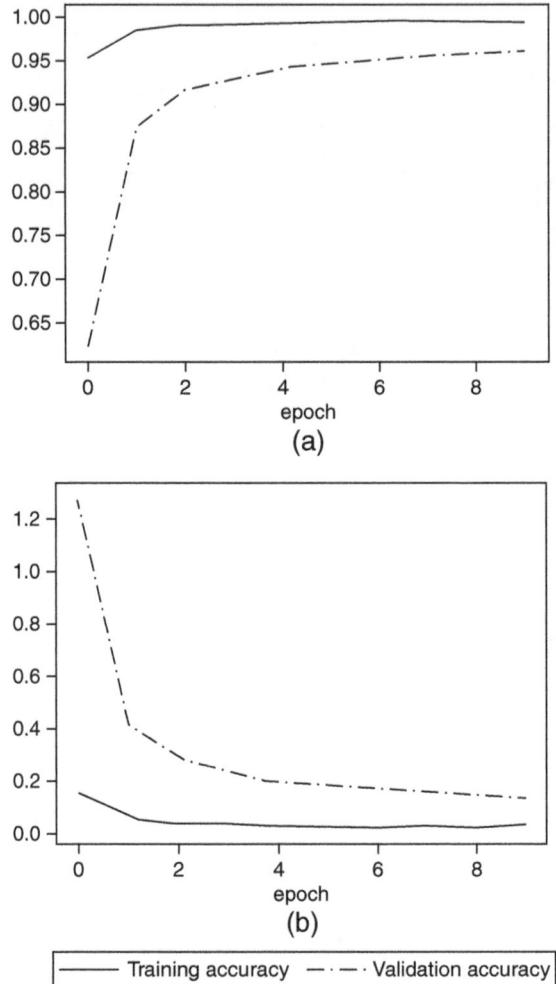

FIGURE 12.10   The CNN model: *(a)* accuracy, and *(b)* loss

3. A HOG sub-sampling (Uijlings et al. 2014) is performed to look inside a small part of the images. This is performed in different scaling factors by referencing the perspective view of the vehicle.
4. The small image extracted is pre-processed and passed through the Deep Learning model already trained in the cloud. This model is adjusted to run inside the GPU of the embedded computer of the AV.

12.4   CONCLUSION

This research presents an AV software architecture, which requires infrastructure in the cloud and the use of convolutional network algorithms with different layers. The

TABLE 12.5
CNN Training Pseudocode

Input parameters: α: Stepsize; Exponential decay rates for the moment estimates; $f(\theta)$: Objective Function (O.F.); θ_0: Initial vector

$m_0 \leftarrow 0$ (Initialize first-moment vector)

$v_0 \leftarrow 0$ (Initialize second-moment vector)

$t \leftarrow 0$ (Initialize timestep)

while θ_t not converged ***do***:

$t \leftarrow t + 1$ (Increase timestep t)

$g_t \leftarrow \nabla_\theta f_t(\theta_{t-1})$ (Get gradients concerning the O.F. at t)

$m_t \leftarrow \beta_1 \cdot m_{t-1} + (1 - \beta_1) \cdot g_t$ (Update biased first-moment estimate)

$v_t \leftarrow v_t/(1 - 2t)$ (Compute bias-corrected second raw moment estimate)

$\theta_t \leftarrow \theta_{t-1} - \alpha \cdot mt/(\sqrt{v}t +)$ (Update)

end while

return θ_t (Outcomes parameters)

Source: Adapted from Kingma and Ba, 2015.

Deep Learning task of the traffic sign classifier was performed with high precision by testing different architectures of these neural networks running within GPU systems in real-time. This Deep Learning model training experiment is conducted on the cloud infrastructure with a discussion of the implication to be used within a fully AV architecture designed to investigate cooperative strategies between the driver and the vehicle's automated system. The results obtained through this application of CNNs are the outcome of a benchmark database experimentation. This research proposes an architecture that unifies the algorithm with AV's hardware allowing us to be closer to real implementations for this type of vehicle that demands greater confidence in future markets.

Future work aims to explore how this tuned Deep Learning algorithm transforms through cloud transfer learning after using the AV-saved data to increase the classification of the new signal. This includes the specific signal of the country where the project is developed. This research opens a new question about how this type of car should be validated in new markets.

REFERENCES

Belcarz, K., T. Białek, M. Komorkiewicz, et al. 2018. "Developing autonomous vehicle research platform – A case study." *IOP Conference Series: Materials Science and Engineering*, 421 (2).

Bishop, C. N. 1995. *Neural Networks for Pattern Recognition.* New York: Oxford University Press.

Bousquet, O. 2012. "The tradeoffs of large scale learning." In *Optimization for Machine Learning*, edited by Survit Sra, Sebastian Nowozin, and Stephen J. Wright. Cambridge, MA: MIT Press: 351–368.

Bouvrie, Jake. 2006. "Notes on Convolutional Neural Networks". Massachusetts Institute of Technology. http://cogprints.org/5869/1/cnn_tutorial.pdf.

Bruno, Diego Renan, Tiago C. Santos, Júnior A. R. Silva, et al. 2018. "Advanced driver assistance system based on automated routines for the benefit of human faults correction in robotics vehicles." *Latin American Robotic Symposium, Brazilian Symposium on Robotics (SBR) and Workshop on Robotics in Education:* 112–117.

Bushaev, V. 2018. "Understanding RMSprop – Faster Neural Network Learning." Towards Data Science. https://towardsdatascience.com/understanding-rmsprop-faster-neural-network-learning-62e116fcf29a.

Cullinane, B., P. Nemec, M. Clement, et al. 2014. "Engaging and Disengaging for Autonomous Driving." United States Patent. https://patents.google.com/patent/US20140156134A1/en.

De Boer, P., D. Kroese, S. Mannor, et al. 2005. "A tutorial on the cross-entropy method." *Annals of Operations Research,* 134 (1): 19–67.

Elhassouny, A. and F. Smarandache. 2019. "Trends in deep convolutional neural network architectures: A review." *2019 International Conference of Computer Science and Renewable Energies:* 1–8.

Garcia, Olmer and Cesar Diaz. 2018. "Machine Learning applied to autonomous vehicles." In *Artificial Intelligence: Advances in research and applications*, edited by L. Rabelo, S. Bhide, and E. Gutierrez. New York: NOVA Science.

Garcia, Olmer, Giovani Bernardes Vitor, Janito Vaqueiro Ferreira, et al. 2018. "The VILMA intelligent vehicle: An architectural design for cooperative control between driver and automated system." *Journal of Modern Transportation,* 26: 220–229.

Geiger, A., M. Lauer, F. Moosmann, et al. 2012. "Team AnnieWAY's entry to the 2011 Grand Cooperative Driving Challenge." *IEEE Transactions on Intelligent Transportation Systems,* 13 (3): 1008–1017.

Goodfellow, I., Y. Bengio, and A. Courville. 2016. *Deep Learning.* Cambridge, MA: MIT Press.

Hassan, Murtaza. 2020. "Murtaza's Workshop." www.murtazahassan.com/courses/opencv-projects/lesson/code-5/.

Heaton, J. 2013. *Artificial Intelligence for Humans, Volume 1: Fundamental algorithms.* Chesterfield, MO: Heaton Research Inc.

Karpathy, A. 2017. "A Peek at Trends in Machine Learning." https://medium.com/@karpathy/a-peek-at-trends-in-machine-learning-ab8a1085a106.

Kato, Shinpei, Eijiro Takeuchi, Yoshio Ishiguro, et al. 2015. "An open approach to autonomous vehicles." *IEEE Micro,* 35: 60–68.

Kingma, Diederik and Jimmy Ba. 2015. "Adam: A Method for Stochastic Optimization." 3rd International Conference for Learning Representations, San Diego. https://arxiv.org/pdf/1412.6980.pdf.

LeCun, Y., L. Bottou, Y. Bengio, et al. 1998. "Gradient-based learning applied to document recognition." *Proceedings of the IEEE,* 86 (11): 2278–2324.

Marin-Plaza, Pablo, Ahmed Hussein, David Martin, et al. 2018. "Global and local path planning study in a ROS-based research platform for autonomous vehicles." *Journal of Advanced Transportation,* 2018: 1–10.

Mosca, Alan and George Magoulas. 2015. "Adapting resilient propagation for Deep Learning." *UK Workshop on Computational Intelligence:* 1–4.

NHTSA (National Highway Traffic Safety Administration). 2013. "Technical Report." www.nhtsa.gov/staticfiles/nti/distracted_driving/pdf/distracted_guidelines-FR_04232013.pdf.

Park, J., B. Bae, J. Lee, at al. 2010. "Design of failsafe architecture for unmanned ground vehicles." *Control Automation and Systems:* 1101–1104.

Passalis, N., A. Tefas, J. Kanniainen, et al. 2020. "Adaptive normalization for forecasting limit order book data using convolutional neural networks." *International Conference on Acoustics, Speech and Signal Processing:* 1713–1717.

Quigley, Morgan, Ken Conley, Brian Gerkey, et al. 2009. "ROS: An open-source robot operating system." *ICRA Workshop On Open Source Software,* 5.

Schwarting, Wilko, Javier Alonso-Mora, and Daniela Rus. 2018. "Planning and decision-making for autonomous vehicles." *Annual Review of Control, Robotics, and Autonomous Systems,* 1: 187–210.

Serre, T., M. Kouh, C. Cadieu, et al. 2005. "A Theory of Object Recognition: Computations and circuits in the feedforward path of the ventral stream in primate visual cortex." Massachusetts Institute of Technology. https://serre-lab.clps.brown.edu/wp-content/uploads/2012/08/GetTRDoc.pdf.

Stallkamp, Johannes, Marc Schlipsing, Jan Salmen, et al. 2020. "Electronic Research Data Archive – University of Copenhagen." https://sid.erda.dk/public/archives/daaeac0d7ce1152aea9b61d9f1e19370/published-archive.html (accessed November 8, 2020).

Uijlings, J. R., I. C. Duta, N. Rostamzadeh, et al. 2014. "Realtime video classification using dense HOF/HOG." *Proceedings of International Conference on Multimedia Retrieval:* 145–152.

Vapnik, V. 2000. *The Nature of Statistical Learning Theory.* Berlin: Springer-Verlag.

Wilson, Ashia, Rebecca Roelofs, Nathan Srebro, et al. 2017. "The Marginal Value of Adaptive Gradient Methods in Machine Learning." https://arxiv.org/abs/1705.08292.

Yao, P., H. Wu, and B. Gao. 2020. "Fully hardware-implemented memristor convolutional neural network." *Nature,* 577: 641–646.

Zhu, D., H. Yao, B. Jiang, et al. 2018. "Negative Log Likelihood Ratio Loss for Deep Neural Network Classification." https://arxiv.org/abs/1804.10690.

13 Optimizing Supply Chain Networks for Specialty Coffee

Santiago Botero López, Muhammad Salman Chaudhry, and Cansu Tayaksi

13.1 THE COFFEE INDUSTRY AND SOCIO-ECONOMIC COSTS FOR COFFEE FARMERS

The coffee industry has grown to be one of the biggest industries in the world, with around US$200 billion annual revenues (Samper, Giovannucci, and Marques 2017). In comparison to the slow-growing food and beverages market, coffee sales are estimated to be at a compounded annual growth rate (CAGR) of 5.5% from the year 2019 to 2024 (Mordor Intelligence 2019). Every year, around the world, people consume around 400 billion cups of coffee and at least 60 million people are economically dependent on it (Sachs et al. 2019).

Coffee production supports the livelihoods of more than 26 million coffee farmers in 52 countries (Hirons et al. 2018). But despite the big population of coffee farmers, their income is estimated at only 2% to 17% of the retail price of standard coffee (UNCTAD 2016), and furthermore, only 1% to 2% of the price paid for a cup of specialty coffee goes to the farmers (Sachs et al. 2019). Day by day, an increasing proportion of farmers fall under the poverty line with their daily incomes becoming under US$1.90 (ICO 2019a).

Today, the consolidation of industry leaders of the coffee industry creates an unbalanced power in the market. While the big players are capturing a large proportion of the price value of the product, in 2019, one-third of Colombian producers were unable to cover their cash outlays and 53% of them operated at a loss when the full costs of production are considered (ICO 2019b). Meanwhile, some industry leaders continued to report their operating profit margins as above 15% (Macrotrends 2019). Thus, at a global level, the gap between the income of producers and retailers continues to widen over time, resulting in a socio-economic crisis for farmers. The situation that coffee farmers are facing brings the need for a transformation in the coffee supply chains. An active step that coffee farmers can take to improve their incomes is to pursue better involvement in the downstream supply chain to gain a higher share of the final price. This participation can be achieved through partnerships with entities that provide downstream services or utilization of new technologies to transform their business models. One of the main opportunities for farmers to transform their

DOI: 10.1201/9781003137993-14

217

business is to sell directly to consumers, retailers, or distributors in consuming countries (Sachs et al. 2019).

Considering the struggling situation of the coffee farmers, we designed a mathematical model to minimize the coffee supply chain costs, from the farms to the distributors in consuming countries. We propose that adopting our model will provide coffee farmers decreased overall costs and therefore increased incomes when selling directly to distributors of roasted coffee.

13.2 COFFEE SUPPLY CHAINS AND A REGIONAL LOOK AT CALDAS, COLOMBIA

Coffee supply chains are complex networks composed of various entities, both in producing and consuming countries. In a typical coffee supply chain, there are entities that transform the product itself and there are entities that do not contain any transformation process. The main entities that transform coffee are: (1) *farms*, where the coffee beans are cropped, pulped, and dried; (2) *mills*, where the husk is removed from the coffee beans; and (3) *roasters*, where the coffee beans are heated and transformed into the aromatic brown roasted coffee beans that are packed and sold to customers. While farms and mills are located in the producing countries, roasters are typically placed in the consuming countries. In addition to the coffee transforming activities, various entities add value to the coffee supply chain by performing activities like trading, selling, marketing, and branding, but they do not transform the product itself. These entities are mainly: (4) *exporters*, who buy coffee in the exporting countries and sell it to entities in other countries; (5) *importers*, who buy coffee from entities in producing countries and sell it to other entities in consuming countries; and (6) *retailers* and *distributors*, who buy coffee from roasters and sell it to end-customers. These entities establish the backbone of the coffee supply chain and each of them either adds value either to the production process or the transportation of coffee.

13.2.1 IMPACT OF THE COFFEE PRODUCTION CHARACTERISTICS ON THE SUPPLY CHAIN

There are 3 main stages of coffee under production. The initial state is called cherry coffee, which is picked by farmers and processed immediately at the farms due to its very short quality stability. The second state is called green coffee, which can be stored without significant quality losses for at least 12 months (Meira Borém et al. 2013) and up to 15 months (Tripetch and Borompichaichartkul 2019). The third state is roasted coffee, which is a result of the roasting process (Bee et al. 2005).

The characteristics of the production process impact mainly on two aspects of the coffee supply chain: the location of facilities and the time constraints. The location of facilities is relevant because the total weight and volume of the coffee beans are impacted after each of the transformation operations (Hicks 2002). These weight and volume variations are mainly due to the removal of the layers that cover the coffee bean, the loss of water, and the physical transformation when exposed to the high temperatures of roasting (Mutua 2000). The second aspect involves time constraints

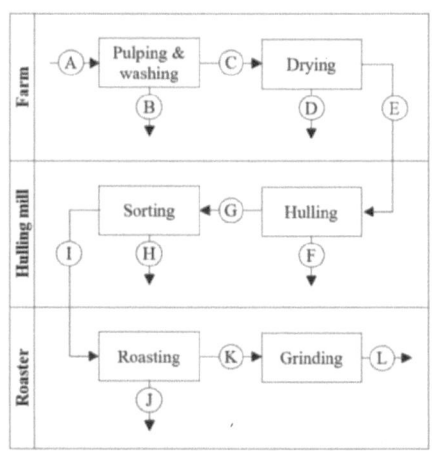

Physical properties of coffee

Product		Weight yields (kg)	Volume (kg/lt)	Stability (months)
A	Cherry coffee	6.88	0.80	0
B	Pulp & mucilage	-	-	-
C	Wet coffee	2.82	0.80	0
D	Water	-	-	-
E	Parchment coffee	1.50	0.40	12–15
F	Husk & silver skin	-	-	-
G	Green coffee	1.25	0.66	12–15
H	Defective coffee	-	-	-
I	Excelso green coffee	1.16	0.66	12–15
J	Water	-	-	-
K	Roasted coffee	1.00	0.37	2–10.6
L	Ground coffee	1.00	0.31	2–10.6

FIGURE 13.1   Stages of coffee production and the physical properties of coffee

due to the stability of the product's quality. Coffee is not a highly perishable product, but its quality is affected through time (Bladyka 2013), and the rate at which it loses quality depends on its state.

The case of roasted coffee is more complex in terms of freshness and shelf life. One of the most determinant factors in slowing the loss of freshness is the packaging technology (Bladyka 2013). Previous research focused on Colombian specialty coffee suggests that experts can perceive a loss of quality in only 10 days, but regular customers do not significantly notice it even 60 days after roasting when packed in traditional bags (Bladyka 2013). Other studies have shown that the use of high barrier aluminum packaging and controlled atmospheres can extend the product's shelf life for up to 46 weeks without significantly affecting the quality (Glöss et al. 2014).

Figure 13.1 shows the stages of coffee production, starting from the moment coffee is picked at farms until it is ready to be sold to consumers. The figure also indicates the typical location at which each of the operations is performed (Bee et al. 2005). The table in the figure lists the weight yields and density at each step of the process (Mutua 2000), and the time coffee can last without significantly losing its quality in the eye of the final customer.

13.2.2 Shipping Coffee Overseas from Caldas, Colombia

The structure of the coffee supply chain is steered by overseas transportation. To understand the existing channels for international shipments of coffee, we analyzed data from The Universidad Nacional de Colombia which includes every coffee shipment exported from the region of Caldas, Colombia, between 2009 and 2018, totaling 28,086 shipments. Caldas is one of the main coffee-producing regions in Colombia with more than US$500 million of yearly exports, 33,000 coffee producers and over 300,000 people depending on the coffee economy (Comité de Cafeteros de Caldas 2019). We focused our research on the shipment behavior between 2009 and

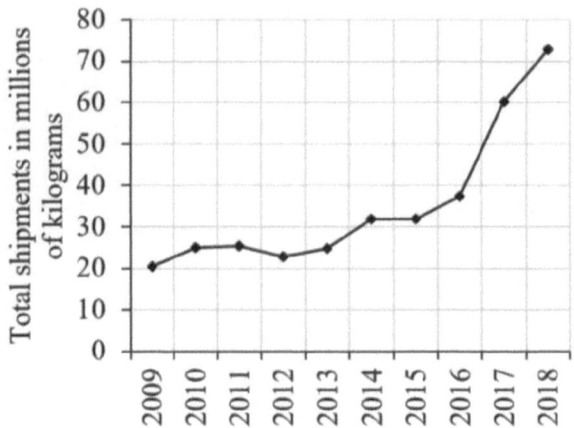

FIGURE 13.2   Yearly shipments of coffee from Caldas to the U.S. (2009–2018)

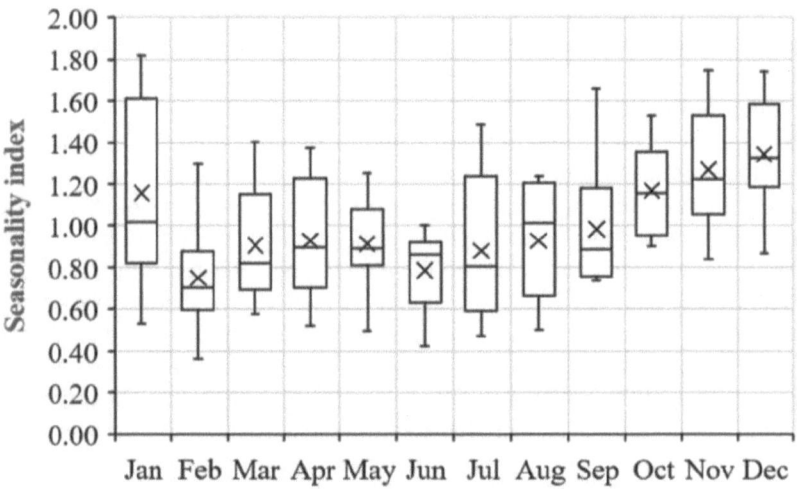

FIGURE 13.3   Seasonal index of yearly coffee exports per month (2009–2018)

2018, transportation channels, and size of shipments and transportation modes, in order to better understand the current demand behavior and the transportation options.

Shipment and seasonality behavior. Figure 13.2 shows the yearly coffee shipments from Caldas to the U.S., which present an upward trend between 2009 and 2018. Strikingly, in the last 2 years of the data (2016–2018), total shipments to the U.S. increased by 190%. In terms of the seasonality of the exports, the monthly seasonality index (Lembke 2015) shows a positive trend throughout the second semester of the year, with the average peaking in December and hitting its lowest point in February. Figure 13.3 shows the seasonal index of yearly coffee exports per month.

TABLE 13.1
Caldas' Coffee Exports by Exporting Port and Type of Product (2009 to 2018)

Departure (Colombia)	Routes	Percentage per Product	
		Green	Roasted
Buenaventura	Sea (Pacific)	56.06%	54.97%
Cartagena	Sea (Atlantic)	39.97%	0.00%
Santa Marta	Sea (Atlantic)	3.96%	0.00%
Bogotá	Air	0.01%	35.52%
Cali	Air	0.00%	7.26%
Medellin	Air	0.00%	2.25%
Total		**100%**	**100%**

Transportation routes. From Caldas, maritime transportation is the primary option for coffee shipping. Almost all (99.9%) the green coffee transported from Caldas to the U.S. is shipped by sea. However, only 54.97% of roasted coffee is shipped by sea, and the rest is transported by air. Table 13.1 shows Caldas' coffee exports by exporting port and type of product.

Shipment sizes and transportation modes. We define the cases in which green coffee was shipped by air or in consolidated pallets as exceptions and we exclude these options for green coffee in this chapter. The shipments' data indicates that the product is predominantly shipped in full container loads (FCL) of approximately 19 tons. Roasted coffee, however, is most commonly transported by air for quantities under 800 kg; and by sea, in consolidated pallets of approximately 900 kg each (up to 5,400 kg), and in FCLs of approximately 19 tons. Figure 13.4 shows the weight of shipments by transportation mode and product type.

13.3   STRUCTURING THE COFFEE SUPPLY CHAIN NETWORK

The characterization of the coffee supply chain starts from mapping the network structure considering the facility types, transportation modes, and the constraints of the processes and product.

Facilities. There are six main types of facilities within the coffee supply chain. Table 13.2 shows each type of facility and the product input and output, the weight and density variations that occur, and the stability of the product after leaving the facilities. For this case, stability does not refer to the shelf life, but to the time it takes for the product to significantly lower in quality in the eye of the final consumer.

Transportation modes. Our model includes multiple transportation modes since the choice between them determines the whole supply chain configuration. The difference in costs, capacities, and transit times affect the decisions on where to locate facilities and where to carry out transformation processes, particularly when considering the product's weight and density variations and its stability. Table 13.3 shows various transportation modes that can be used at different stages along the supply chain.

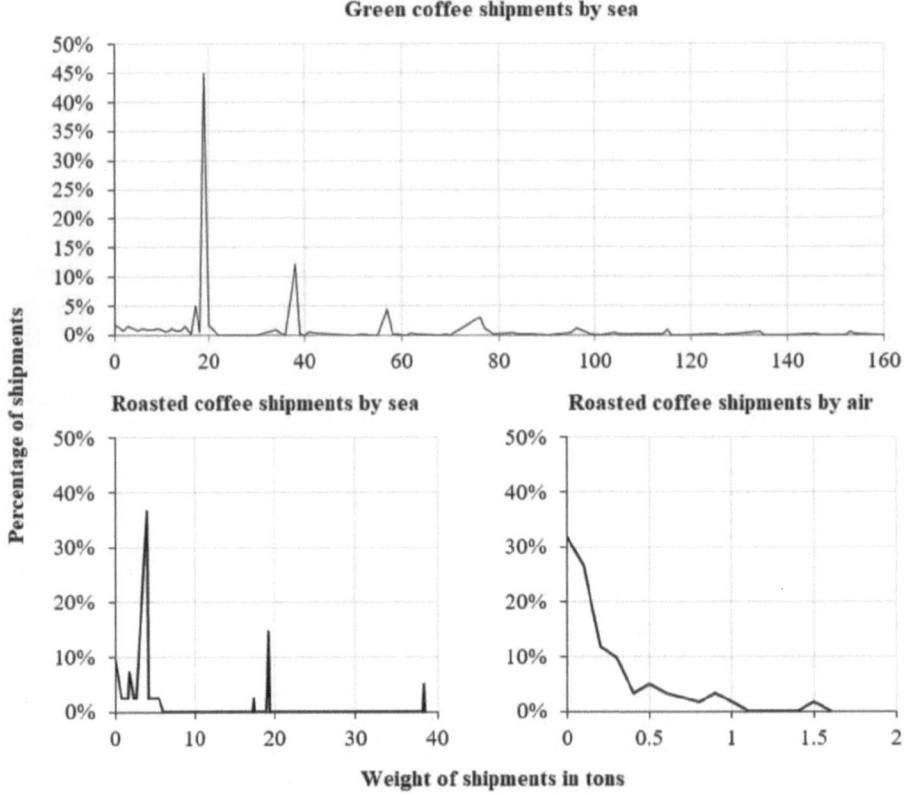

FIGURE 13.4   Weight of shipments by transportation mode and product type

TABLE 13.2
Types of Facilities in the Coffee Supply Chain

Facility Type	Notation	Coffee Input	Coffee Output	Weight Variation	Density Variation	Stability (months)
Farm	s	N/A	Parchment	-22.67%	65.00%	N/A
Mill	m	Parchment	Green	-22.67%	65.00%	12
Roaster – T1	r(t1)	Green	Roasted	-17.50%	-48.48%	2
Roaster – T2	r(t2)	Green	Roasted	-17.50%	-48.48%	10.6
Warehouse	w	Green/Roasted	Green/Roasted	0%	0%	N/A
Distributor	c	Roasted	N/A	0%	0%	N/A

Note: T1: traditional packing technology; T2: controlled atmosphere packing technology.

TABLE 13.3
Transportation Modes in the Coffee Supply Chain

Transportation Mode	Notation	Applicable Products	Applicable Region	Shipping Unit	Transit Time (days)
Sea – FEU (40 ft. container)	f	Green/Roasted	International	FEU	15–25
Sea – TEU (20 ft. container)	t	Green/Roasted	International	TEU	15–25
Sea –Consolidated pallet	p	Roasted	International	Pallet	15–25
Air	a	Roasted	International	kg	5–10
Land – Bulk	b	Parchment/Green	Colombia	kg	1–2
Land – Unit	d	Green/Roasted	United States	kg	1–2

Note: FEU: forty-foot equivalent unit; TEU: twenty-foot equivalent unit.

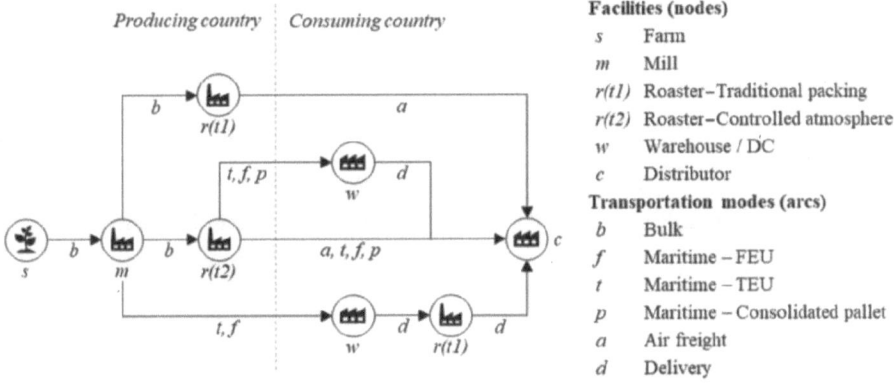

FIGURE 13.5 Simplified supply chain network diagram

Figure 13.5 shows a simplified network diagram that considers the possible routes of the supply chain. In the network, every route complies with the supply chain's constraints. Thereby, we ensure the product is delivered to the customer within its stability time constraint. The figure follows the notations presented in Tables 13.2 and 13.3, and the facilities and transportation modes are also listed.

13.3.1 SUPPLY CHAIN NETWORK DESIGN

Our supply chain network design (SCND) model is a mixed-integer linear program (MILP) model. SCND models are characterized by the following properties (Bertsekas 1998):

- A set of m supply nodes. Supply node i can supply at most S_i units.
- A set of n demand nodes. Demand nodes j must receive at least D_j units of the shipped goods.
- Each unit produced at supply point i and shipped to demand point j incurs a variable cost c_{ij}
- S and D are the supply and demand constraints.
- M_{ij} is an arbitrary large number, specific to each arc (the value could be similar between arcs).
- There are linking constraints $\forall ij$ to ensure we do not allocate shipments to a location that is not used.
- P_{min} and P_{max} are constraints on the number of facilities to use. The sum of the Y-variables will be the total number of facilities in use.
- There are non-negativity constraints $\forall ij$ and $\forall i$, for the X and Y variables.

The objective of our SCND model is to find the path that minimizes cost while respecting the model's constraints. The general model is as follows:

Minimize

$$z = \sum_i \sum_j c_{ij} X_{ij} + \sum_i f_i Y_i$$

Subject to,

$$\sum_j x_{ij} \leq s_i \qquad \forall i \in S \qquad \text{(supply constraints)}$$

$$\sum_i x_{ij} \geq Dj \qquad \forall j \in D \qquad \text{(demand constraints)}$$

$$x_{ij} - M_{ij} Y_i \leq 0 \qquad \forall ij \qquad \text{(linking constraints)}$$

$$\sum_i Y_i \geq P_{min} \qquad\qquad\qquad \text{(minimum facilities constraint)}$$

$$\sum_i Y_i \leq P_{max} \qquad\qquad\qquad \text{(maximum facilities constraint)}$$

$$x_i \geq 0 \qquad \forall ij \qquad \text{(non-negativity constraint)}$$

$$Y_i = \{1\} \qquad \forall i \qquad \text{(binary constraint)}$$

Once formulated, the SCND model's parameters are loaded with data for transportation (arcs) and facilities (nodes). Transportation data includes cost and capacities associated with both inbound and outbound transportation. Facilities data includes fixed and variable costs of the facilities and their capacity.

13.3.1.1   Model Formulation

In this section, we present the formulation of our SCND model. The network consists of all the candidate nodes and arcs considered by the proposed model, and the objective function finds the feasible route of least cost from the supply nodes to the demand nodes, subject to all the constraints.

The candidate locations are all the facilities that can be used within the supply chain distribution network. Multiple locations are considered for each facility type, and multiple transportation modes can be considered between every pair of nodes. In the case of roasting facilities, packing technologies (traditional and controlled atmosphere) will be modeled as variable indexes, so each node has the option to adopt either of the two.

To define our SCND model, we use the following *decision variables*:

y_f^p	binary variable to determine if candidate facility f operates on period p
x_f^p	incoming flow of product to facility f on period p, in kilograms
ry_{rek}^p	binary variable to determine if roasting facility r operates using technology e and capacity k on period p
rx_{rek}^p	incoming flow of product to roasting facility r with technology e and capacity k on period p, in kilograms
ty_{ijt}^p	binary variable to determine if product flows from nodes $i \in N$ to $j \in N$ on transportation mode t, on period p
tx_{ijt}^p	flow of product from node $i \in N$ to node $j \in N$ on the transportation mode t, on period p, in kilograms
tz_{ijt}^p	integer number of shipping units that flow from node $i \in N$ to node $j \in N$ on the transportation mode t, on period p

After defining the decision variables, we continue with all the *parameters* of the model:

f_f^p	fixed cost of opening the candidate facility f on period p
c_f^p	cost of transforming or storing one kilogram of product at candidate facility f on period p
rf_{rek}^p	fixed cost of opening roasting facility r with technology e and capacity level k, on period p
rc_{rek}^p	cost of roasting one kilogram of green coffee at facility r with technology e and capacity k, on period p
tf_{ijt}^p	fixed cost of transporting product from nodes $i \in N$ to $j \in N$ on transport mode t, on period p
tc_{ijt}^p	cost of transporting one kilogram of product from nodes $i \in N$ to $j \in N$ on transport mode t, on period p
tq_{ijt}^p	cost of transporting one shipping unit from nodes $i \in N$ to $j \in N$ on transport mode t, on period p
dem_c^p	demand for roasted coffee on node c, on period p, in kilograms
M	constant with a very high value used for the binary linking constraints
cap_f	max capacity per period of candidate facility f, in kilograms
$rcap_{rek}$	max capacity per period of roasting facility r with technology e and capacity level k, in kilograms
twc_{ijt}	max weight per period to transport from nodes $i \in N$ to $j \in N$ on transport mode t
tvc_{ijt}	max volume per period to transport from nodes $i \in N$ to $j \in N$ on transport mode t, in liters
suw_t	max capacity of product per shipping unit for transportation mode t, in kilograms
suv_t	max volume of product per shipping unit for transportation mode t, in liters
den_{ij}	density ($kg \,/\, m^3$) of the product being transported from node $i \in N$ to node $j \in N$
wv_n	weight reduction percentage of product in node n

The *index sets* of the SCND model are the following:

P :   set of time periods, $p \in P = \{1,2,3,\ldots,12\}$
T :   set of transportation modes, $t \in T = \{bulk, \; air, \; pallet, \; TEU, \; FEU, \; distribution\}$
E :   set of roasting and packing technologies, $e \in E \{Traditional, \; Controlled \; atmosphere\}$
K :   set of roaster capacities levels (kg/period), $k \in K = \{15000, 42000, 72000, 150000\}$
S :   sourcing farms
C :   demand nodes
M :   candidate milling facilities
W :   candidate warehousing and distribution facilities
R :   candidate roasting and packing facilities
N :   set of all nodes $\{S \cup C \cup M \cup W \cup R\}$
F :   set of non-roasting nodes $\{S \cup C \cup M \cup W\}$

Where the objective function is:

$$
\text{Minimize} \sum_{p \in P} \left(\begin{array}{l} \displaystyle\sum_{f \in F} \left(y_f^p \; f_f^p + x_f^p \; c_f^p \right) + \sum_{r \in R} \sum_{e \in E} \sum_{k \in K} \left(ry_{rek}^p \; rf_{rek}^p + rx_{rek}^p \; rc_{rek}^p \right) \\ + \displaystyle\sum_{i \in N} \sum_{j \in N} \sum_{t \in T} \left(ty_{ijt}^p \; tf_{ijt}^p + tx_{ijt}^p \; tc_{ijt}^p + tz_{ijt}^p \; tq_{ijt}^p \right) \end{array} \right) \quad (1)
$$

The first two terms of the expression in brackets define the facilities. It includes both fixed and variable costs of operating the two types of facilities which are non-roasting facilities (first bracket) and roasting facilities (second bracket). The third term in the expression defines the transportation. It includes the fixed costs of using a transportation mode, the variable cost per kilogram, and the variable cost per shipping unit of transportation.

The constraints of the model are:
Subject to

$$x_c^p \geq dem_c^p \qquad\qquad c \in C, \; p \in P \qquad\qquad\qquad (2)$$

$$x_f^p \leq y_f^p M \qquad\qquad f \in F, p \in P \qquad\qquad\qquad (3)$$

$$rx_{rek}^p \leq ry_{rek}^p \; M \qquad\quad r \in R, e \in E, k \in K, p \in P \qquad (4)$$

$$tc_{ijt}^p \leq ty_{ijt}^p M \qquad\quad i \in N, j \in N, t \in T, \; p \in P \qquad (5)$$

$$x_f^p \leq cap_f \qquad\qquad\quad f \in F, p \in P \qquad\qquad\qquad (6)$$

$$rx_{rek}^p \leq rcap_{rek} \qquad\quad r \in R, \; e \in E, \; k \in K, \; p \in P \qquad (7)$$

$$tx_{ijt}^p \le tcap_{ijt} \qquad\qquad i \in N, j \in N, t \in T, p \in P \qquad (8)$$

$$tz_{ijt}^p \ge \frac{tx_{ijt}^p}{suw_t} \qquad\qquad i \in N, j \in N, t \in T, p \in P \qquad (9)$$

$$tz_{ijt}^p \ge \frac{tx_{ijt}^p \, den_{ij}}{suv_t} \qquad\qquad i \in N, j \in N, t \in T, p \in P \qquad (10)$$

$$x_j^p \left(1 + wv_n\right) \ge \sum_{i \in N} \sum_{t \in T} \left(tx_{jit}^p\right) \qquad j \in M, j \in R, t \in T, p \in P \qquad (11)$$

$$x_j^p \le \sum_{i \in N} \sum_{t \in T} \left(tx_{ijt}^p\right) \qquad j \in M, j \in R, t \in T, p \in P \qquad (12)$$

$$x_s^p - \sum_{i \in I} \sum_{t \in T} tx_{sit}^p = 0 \qquad s \in S, p \in P \qquad (13)$$

$$x_c^p - \sum_{i \in I} \sum_{t \in T} tx_{ict}^p = 0 \qquad c \in C, p \in P \qquad (14)$$

$$y_i^p \ge y_i^{p-1} \qquad\qquad i \in N, p \in P \qquad (15)$$

$$y_f^p \quad binary \qquad\qquad f \in F, p \in P \qquad (16)$$

$$ry_{rek}^p \quad binary \qquad\qquad r \in R, e \in E, k \in K, p \in P \qquad (17)$$

$$ty_{ijt}^p \quad binary \qquad\qquad i \in N, j \in N, t \in T, p \in P \qquad (18)$$

$$tz_{ijt}^p \quad integer \qquad\qquad i \in N, j \in N, t \in T, p \in P \qquad (19)$$

$$x_f^p \ge 0 \qquad\qquad f \in F, p \in P \qquad (20)$$

$$rx_{rek}^p \ge 0 \qquad\qquad r \in R, e \in E, k \in K, p \in P \qquad (21)$$

$$tx_{ijt}^p \ge 0 \qquad\qquad i \in N, j \in N, t \in T, p \in P \qquad (22)$$

$$tq_{ijt}^p \ge 0 \qquad\qquad i \in N, j \in N, t \in T, p \in P \qquad (23)$$

The constraints above represent the area in which the MILP has the flexibility to find an optimal solution, satisfying the requirements for the customer, suppliers, and logistics capacities.

Equation 2 guarantees that the demand of every customer d is fulfilled for every time period. Constraints 3, 4, and 5 ensure that whenever a product flows through a node or arc, the binary variable associated takes a value of 1. This forces the model to consider the fixed costs incurred when utilizing nodes and arcs.

Constraints 6, 7, and 8 maintain the utilization levels of every node and arc equal to or below its maximum capacity.

Constraint 9 ensures that the necessary amount of shipping units are used for each arc based on the product's weight, for every transportation mode. It does so by forcing the shipping units to take an equal or higher integer value than the total amount of product in kilograms divided by the shipping unit's weight capacity.

Constraint 10 ensures that the necessary amount of shipping units are used for each arc based on the product's volume, for every transportation mode. It does so by forcing the number of shipping units to take an equal or higher integer value than the total amount of product in liters divided by the shipping unit's volume capacity. The product's volume is calculated by multiplying its weight and density in kilograms/ liters.

Constraints 11 to 14 guarantee network continuity. Equations 11 and 12 imply that the total amount of product in kilograms that goes into a node is equal to the amount that goes out plus the waste. The waste is calculated by multiplying the total weight input and the weight variation factor. This only applies to nodes that transform coffee (mills m and roasters r).

Constraints 13 and 14 create exceptions for farms and customer nodes, to ensure that they behave as sources and sinks. Equation 13 guarantees that the amount of product going out of each supply node through transportation arcs is equal to the amount of product that the node supplies. Constraint 14 implies that the amount of product going into every demand node through transportation arcs is equal to the amount of product that the node demands.

Equation 15 ensures that whenever a node is opened, it remains open for every remaining period. This is particularly important for the facilities that imply investment to open.

Constraints 16, 17, and 18 declare the binary variables used to calculate the fixed cost of opening nodes and arcs. Constraint 19 declares the integer variable that counts the shipping units flowing through every arc for each transportation mode. Equations 20 to 23 represent the variables' non-negativity constraints.

13.3.2 Validating with a Case from Colombia: Café Botero

The Botero family has been growing coffee in Caldas, Colombia, since 1907; and like most farmers worldwide, they have been impacted by the uneven distribution of the industry's revenue in recent years.

In 2013, the family decided to take active steps to pursue more participation in the downstream supply chain, to gain a higher share of the final price. Following this decision, they founded a new company, called Café Botero, dedicated to processing and commercializing roasted coffee as a finished product. Until 2020, Café Botero has focused on the Colombian business-to-business (B2B) market of specialty roasted coffee, selling to over 250 customers since 2013.

Café Botero has explored markets abroad, carrying out limited international shipments of roasted coffee to the U.S. However, the company lacks a structured supply chain for these international shipments. This situation created an opportunity for us to apply our model and optimize the supply chain of the company, minimizing the total cost from their farms in Caldas, Colombia, to their customers in the U.S.

13.3.2.1 Validation Scenarios

We validated our model through a prediction experiment, where we fixed the parameters of the model at historic real values and compared the results with the real costs of the company's supply chain (McCarl and Apland 1986). For this prediction experiment we replicated previous roasted coffee shipments by Café Botero to the U.S. in 2018. The total cost per kilogram calculated by the model was 3.63% lower than the real observed cost per kilogram of the historic shipments. We used this result as the baseline scenario for the following step.

To assess the robustness of our model, we developed eight demand scenarios: four demand states as the Stable, Growing, Seasonal, and Growing Seasonal demands; and 2 sub-scenarios for each of them: low level of demand and high level of demand. Figure 13.6 shows the demand scenarios. For each of them, we optimized the supply chain network.

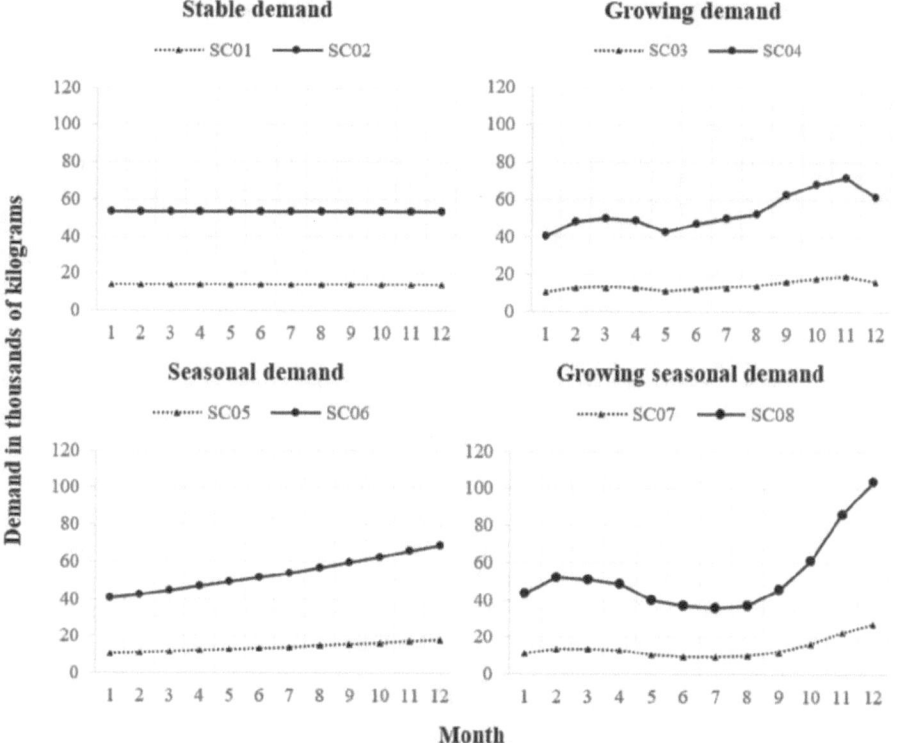

FIGURE 13.6 Demand scenarios: stable, growing, seasonal, and growing seasonal

TABLE 13.4
Savings for Each Demand Scenario

Scenario	Demand Level	Monthly Growth	Demand Seasonality	Savings		
				Arcs	Nodes	Total
SC01	Low	No	No	57.66%	14.78%	52.42%
SC02	High	No	No	83.54%	46.00%	78.96%
SC03	Low	No	Yes	55.16%	19.58%	50.82%
SC04	High	No	Yes	83.68%	45.56%	79.03%
SC05	Low	Yes	No	54.37%	19.24%	50.08%
SC06	High	Yes	No	83.55%	45.73%	78.93%
SC07	Low	Yes	Yes	55.66%	19.58%	51.26%
SC08	High	Yes	Yes	83.09%	38.47%	77.64%

13.3.2.2   Results of the Scenarios and Saving Opportunities

By running the optimization model, we calculated the best supply chain configuration for each of the eight scenarios. Table 13.4 shows the savings for each demand scenario on transportation (arcs), at production facilities (nodes), and the supply chain totals.

Our model considers multiple decisions such as the specific location of the production and storage facilities, their capacity, and the specific supply chain configuration for each month throughout the year. As expected, there were different supply chain configurations for each of the scenarios and even variations throughout the 12 months.

These differences in the optimal configuration are due to the demand levels, seasonality, and other specifics of each scenario. For example, for scenarios of higher demand, the model recommends having international shipping on full container loads, while lower demand scenarios should use consolidated pallets. The required capacity of the production facilities also changes depending on the total volume of coffee to be roasted.

However, the general structure remained the same in every case. For every scenario, the optimal configuration considers doing the milling and roasting processes at the same location, as close as possible to the farms. Neither of these processes should be outsourced (except for scenarios in which the demand seasonality generates high demand peaks, when a portion of the production should be outsourced). The roasting facilities should use the controlled atmosphere packing technology to extend the product's quality stability and enable longer transportation times. From the roasting facility, coffee is shipped directly to distributors in the consuming country using TEUs (twenty-foot equivalent units) or consolidated pallets, depending on the size of the shipments. See Figure 13.7 for this general optimal supply chain configuration.

13.3.2.3   Recommendations for Café Botero

Our model generates multiple recommendations for Café Botero. The first recommendation is to minimize the number of facilities in the supply chain and place

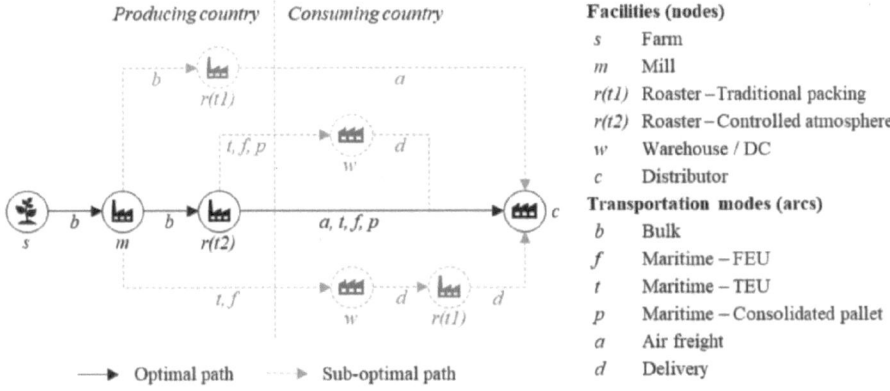

FIGURE 13.7   General optimal supply chain configuration

them in Caldas, Colombia. This generates cost savings due to: (1) the lower costs of operating facilities in Colombia compared to the U.S.; (2) the reduction in the total number of facilities, eliminating unnecessary transportation and handling costs; and (3) the minimization of the total transported weight generated by processing coffee closer to the farms.

Second, leveraging economies of scale whenever possible brings significant savings in the total supply chain, as the scenario analysis demonstrated great savings in the total supply chain cost per kilogram when the throughput increases.

The third recommendation is to package the roasted coffee utilizing controlled atmosphere technologies to prolong its shelf life and its quality stability, allowing the shipment of coffee by sea and hence generating savings in transportation.

Lastly, we recommend that the company re-runs the model as more and better data is available. This iterative process will generate continuous improvement to both the company's supply chain and the model itself.

13.4   ACTIVE STEPS DOWN THE SUPPLY CHAIN TO REDUCE COSTS

The results of our analysis let us conclude that engineering analytics models can be developed and applied to improve the supply chain of coffee. These models can be used not only to reduce costs and increase the efficiency of current systems, but also to support farmers in their effort to add and capture more of the industry's total value through efficient vertical integration.

By applying our model to eight different demand scenarios at Café Botero, we identified relevant trends that could significantly reduce costs in the supply chain of coffee. First, roasting the coffee in the producing countries (as close to the farms as possible) reduces the costs of production and transportation. Second, using roasted coffee packaging technologies that extend the quality stability time will increase opportunities for international shipments by sea and therefore results in reduced transportation costs. Third, to minimize the supply chain cost, it is necessary for farmers to take over the whole production process, including roasting. And last but not least, opportunities coming from the economies of scale in the cost per kilogram are crucial to generate additional competitive advantage; therefore, the adaptation of strategies to increase the total throughput of the supply chain is vital for the coffee farmers.

The optimal supply chain recommended by our model is structurally different from the current typical coffee supply chain. The savings generated by this optimized structure could potentially generate enough competitive advantage for farmers to be successful in taking active steps downstream in the supply chain and capturing more of the product's value. Thus, we conclude that engineering analytics supports the development of new and transformed business models for coffee farmers, helping them to re-balance the distribution of the product's value throughout the value chain.

13.5   AGENDA FOR FUTURE RESEARCH IN COFFEE SUPPLY CHAINS

This chapter highlights the opportunity to improve the coffee supply chain through network design. This is an improvement that has the potential to impact the lives of thousands of families that are economically dependent on coffee production.

Therefore, we propose three areas for further research. First, expanding our model for more or bigger geographies. By covering larger production and consumption territories, our model will be improved to positively impact the income of more producers.

Second, developing similar supply chain network design models for other similar industries. Like coffee, other crops are produced in developing countries where producers could benefit from the efficient vertical integration of supply chains. Designing optimal supply chains for those products could also generate higher income for farmers in other industries that may have similar problems.

Lastly, we recommend extended research in particular aspects of our proposed supply chain network design. Using engineering analytics to further develop the packaging technologies, the production processes within the facilities, and the marketing of the product would help improve and materialize the recommendations proposed by our research.

In this chapter, we have discussed the struggling economic situation that coffee farmers are facing and the opportunity to reduce the supply chain costs for them. By developing a supply chain network design model and a scenario analysis, we present to the reader the cost reduction opportunities for the coffee farmers and open a door for improving their livelihoods. We hope this chapter will contribute to further research in the field and improve the lives of those that are economically dependent on coffee production in many countries.

REFERENCES

Bee, S., C. Brando, G. Brumen, et al. 2005. "The raw bean." In *Espresso Coffee: The science of quality*, edited by A. Illy and R. Viani. San Diego: Elsevier Academic Press: 87–178.
Bertsekas, D. P. 1998. *Network Optimization: Continuous and discrete models*. Belmont, MA: Athena Scientific.
Bladyka, E. 2013. *Coffee Staling Report*. Santa Ana, CA: Specialty Coffee Association of America.

Comité de Cafeteros de Caldas. 2019. "Federación Nacional de Cafeteros de Colombia." https://caldad.federaciondecafeteros.org/fnc/nuestros_cafeteros/category/118 (accessed October 20, 2019).

Glöss, A. N., B. Schönbächler, M. Rast, et al. 2014. "Freshness indices of roasted coffee: Monitoring the loss of freshness for single serve capsules and roasted whole beans in different packaging." *CHIMIA International Journal for Chemistry,* 68 (3): 179–182.

Hicks, A. 2002. "Post-harvest processing and quality assurance for specialty/organic coffee products." *AU Journal of Technology*, 5.

Hirons, M., Z. Mehrabi, T. A. Gonfa, et al. 2018. "Pursuing climate resilient coffee in Ethiopia – A critical review." *Geoforum*, 91: 108–116.

ICO (International Coffee Organization). 2019a. *Survey on the Impact of Low Coffee Prices on Exporting Countries*. Nairobi, Kenya: International Coffee Council.

ICO (International Coffee Organization). 2019b. *Profitability of Coffee Farming in Selected Latin American Countries – Interim report*. Nairobi, Kenya: International Coffee Council.

Lembke, R. 2015. "Forecasting with Seasonality." University of Nevada. http://business.unr.edu/faculty/ronlembke/handouts/Seasonality%20Final17.pdf.

Macrotrends. 2019. "Starbucks Profit Margin 2006-2019." www.macrotrends.net/stocks/charts/SBUX/starbucks/profit-margins (accessed June 30, 2019).

McCarl, B. A. and J. Apland. 1986. "Validation of linear programming models." *Southern Journal of Agricultural Economics*, 18 (2): 1–10.

Meira Borém, F., F. Carmanini Ribeiro, L. Pereira Figueiredo, et al. 2013. "Evaluation of the sensory and color quality of coffee beans stored in hermetic packaging." *Journal of Stored Products Research*, 52: 1–6.

Mordor Intelligence. 2019. "Coffee Market – Growth, trends and forecasts (2019–2024)." www.reportlinker.com/p05797241/Coffee-Market-Growth-Trends-and-Forecasts.html.

Mutua, J. 2000. *Post-harvest Handling and Processing of Coffee in African Countries*. Rome: Food and Agriculture Organization of the United Nations.

Sachs, J., K. Y. Cordes, J. Rising, et al. 2019. "Ensuring Economic Viability and Sustainability of Coffee Production." Columbia Center on Sustainable Investment. https://ccsi.columbia.edu/content/ensuring-economic-viability-and-sustainability-coffee-production-0.

Samper, L., D. Giovannucci, and L. Marques. 2017. "The powerful role of intangibles in the coffee value chain." World Intellectual Property Organization. *WIPO Economic Research Working Papers,* 39.

Tripetch, P. and C. Borompichaichartkul. 2019. "Effect of packaging materials and storage time on changes of colour, phenolic content, chlorogenic acid and antioxidant activity in arabica green coffee beans." *Journal of Stored Products Research*, 84 (11).

UNCTAD (United Nations Conference on Trade and Development). 2016. "Commodities at a Glance: Special issue on coffee in East Africa." https://unctad.org/webflyer/commodities-glance-special-issue-coffee-east-africa.

14 Spatial Analysis of Fresh Food Retailers in Sabana Centro, Colombia

Agatha da Silva, Daniela Granados-Rivera, Gonzalo Mejía, Christopher Mejía-Argueta, and Jairo Jarrín

14.1 INTRODUCTION

Two factors that favor food security are the ease of physical access to the different food distribution channels and stores available in the retail and food environment, and the improvement of the economic possibilities for the acquisition of food, which enriches individual's nutrition. The latest data available for Latin America and the Caribbean show an increase of 4.5 million people with food insecurity when comparing 2018 to 2014 (FAO, FIDA et al. 2020). Some emerging economies, such as Colombia, are privileged due to the wide production capacity of fresh and healthy food such as fruits and vegetables. However, the same situation does not apply to the food distribution and commercialization channels, which makes its access difficult and expensive, especially for inhabitants of the most vulnerable socio-economic strata.

This chapter focuses on characterizing the accessibility and affordability of fruits and vegetables, spatially and statistically, in the Sabana Centro region of Colombia. This region lies in the high plateau of central Colombia. It encompasses 10 municipalities (Cajicá, Chía, Cogua, Cota, Gachancipá, Nemocón, Sopó, Tabio, Tenjo, and Tocancipá y Zipaquirá), and Chía and Cajicá (see Figure 14.1) are the two most important municipalities due to their proximity to the country's capital, Bogotá. Chía and Cajicá combine modern architecture, good connectivity through multi-lane highways, large shopping malls, and top universities with small town grid layout, a semi-rural environment, and traditional dwellers. The Food Supply Plan of the Central Region of Colombia – Prospective Analysis of Food Supply of the Central Region 2019–2030 shows that the municipalities of the Sabana Centro region are affected by a high level of food insecurity (52.5%) and suggests that this situation is generated by the limited access and availability to food (RAP-E 2021).

Six out of 10 municipalities from the Sabana Centro region have a population greater than 30,000 inhabitants. Actually, Zipaquirá is home to 138,654 inhabitants,

DOI: 10.1201/9781003137993-15

Chía to 141,308, and Cajicá to 87,866 inhabitants, being the largest three municipalities of the region (Sabana Centro Cómo Vamos 2018). These municipalities also display the largest population densities of the region: Chía – 1,859.3 inhabs/km², Cajicá – 1,657.8 inhabs/km², and Zipaquirá – 714.7 inhabs/km². Along the same lines, Cajicá shows the biggest concentration of population in urban areas with 15,575 inhabs/km² followed by Chía with 14,790 inhabs/km². These figures resemble the population density of Bogotá.

The food retail landscape in the two main parts of the municipality encompasses several categories of business, including the traditional small, family-owned stores or *nanostores*, specialized fruit and vegetable retailers (called *fruvers,* locally), as well as small and large supermarkets. This chapter analyzes the relationship between retail coverage, the geographical dispersion of various fresh food distribution channels, and the socio-economic level of individuals and households in two municipalities of Sabana Centro: Chía and Cajicá. We address the following research questions: How are spatial features like location (i.e., convenience), and the geographical dispersion of the retailers, influenced by the socio-economic factors in the municipalities of the Central Sabana region of Colombia? What is the effect on food accessibility and affordability for diverse population segments?

To address these questions, we proposed a methodological framework based on using geographical information systems, to analyze the geographical limitations and the effects of location of multiple retail formats. In addition, we used spatial statistics to derive insights about food accessibility and affordability in Chía and Cajicá using

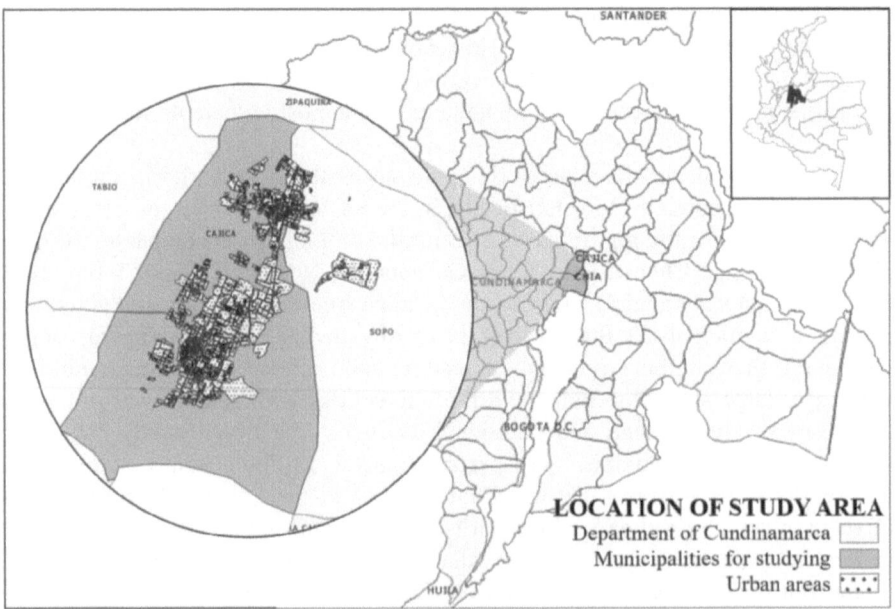

FIGURE 14.1   Location of the area under study within Colombia and the state of Cundinamarca

catchment areas and spatial clustering, to derive the effects of density and proximity to large road infrastructure.

The rest of this chapter is organized as follows: Section 14.2, a literature review, also contains a definition of each type of food retailer. Section 14.3 contains a conceptual presentation about the connection between the commercial panorama, the demographic and socio-economic characteristics, and the geographic characteristics, as well as discussing the methodology used to perform the research. Section 14.4 presents the results of our research, and Section 14.5 contains the conclusions and contributions of this research.

14.2   LITERATURE REVIEW

According to the World Health Organization (2020), malnutrition can be defined as deficiencies, excesses, or imbalances in a person's intake of energy and/or nutrients, leading to undernourishment, overweight, and obesity conditions. This section presents a brief overview of the literature body that justifies the importance of addressing food malnutrition by combining spatial geography and supply chain management perspectives.

First, we introduce important facts related to food malnutrition and how the COVID-19 crisis has exacerbated the need for better supply chains and retailing environments. Second, we discuss the relationship between spatial geography, and retail and supply chains. Third, we position how geography, urban demographic features, and the diversity of retailing impact food accessibility; while socio-economic characteristics (i.e., affordability) and assortment of produce items (i.e., availability) affect demand at multiple retail formats. Finally, we present the practical and theoretical gaps addressed by this chapter.

14.2.1   TRENDS AND FACTS ABOUT FOOD INSECURITY

Around 30% of the world's population suffers from some type of food malnutrition (i.e., food insecurity, overweight, or obesity) (FAO, IFAD et al. 2020). Costs associated with food malnutrition are estimated at around US$4 trillion annually (Development Initiatives Poverty Research Ltd. 2018) due to public health expenditure in prevention and corrective programs, as well as in investing in ways to shape consumer's behavior toward buying and eating healthier products by making them accessible, available, and affordable (Qin et al. 2019). Therefore, better food supply chain strategies are required to increase geographical accessibility at various retail formats and the availability of fresh products like fruits, vegetables, and legumes for all population segments, but especially low-income groups.

Disconnected food retail environments to household needs and preferences, together with poor food supply chains, are still a pivotal problem for our society worldwide (Development Initiatives Poverty Research Ltd. 2020). This has been highlighted by the framework of the United Nations sustainable development goals. However, hunger in the world has kept rising since 2014, with an increase of 60 million people with food insecurity in the last five years. Unfortunately, in Latin America and the Caribbean, which is considered the most unequal region of the world, the situation has also worsened since 2014. Figures have increased from 34.5 million to 47.7 million people living in food-insecure conditions in the region (FAO, IFAD et al. 2020). It is also expected that the COVID-19 pandemic had expanded the number of food-insecure people from developing countries to 265 million by the end of 2020 (Anthem 2020). Thus, to address this growing problem, novel strategies and regulations must be proposed to tackle and control the vicious circle that the lack of affordability and accessibility to nutritious food creates in under-served communities.

Food environments depend heavily on household purchasing patterns, individual food choices, situational factors (i.e., convenience), socio-economic features, and cultural background (Qin et al. 2019; Development Initiatives Poverty Research Ltd. 2020). The best way to shift non-produce-related consumption is by using effective supply chains to connect food supply and production with distribution and diverse markets conveniently and affordably. However, food supply chains in emerging markets are still fragmented (Mejía-Argueta et al. 2019). This fragmentation carries higher logistics complexity, a larger number of intermediaries, and costly last-mile distribution to under-served communities. These effects create inequities due to the non-presence of core food groups (e.g., fruits, vegetables, and legumes) in certain geographical areas or, if they are available, they will be expensive (Mejía-Argueta et al. 2019; Development Initiatives Poverty Research Ltd. 2020) . Thus, vulnerable population segments may suffer from food malnutrition not only because they do not have enough to eat, but also because they eat too much high-calorie content and low-nutrition food (i.e., sugary, or ultra-processed food) instead of costly fresh produce items (FAO, IFAD et al. 2020).

In Latin America and Colombia, healthy diets are 60% more expensive than in other regions of the world for food-insecure people (FAO, IFAD et al. 2020). Hence, when addressing the absence of food variety for end consumers, the income level, the assortment of food at retail formats, the geographic conditions, and the lack of infrastructure may hinder the logistics operations and availability of nutritious food. Therefore, academics, decision- and policymakers must enhance the adequate access to fruits and vegetables or other nutritious food by strengthening supply chains that increase food variety in the retail environment, and by creating alternative food access models and effective strategies to increase availability of healthier assortments, so that the demand for produce items matches the shopping habits of diverse individuals and households (Dannefer et al. 2012; Bauerová 2018; Paluta et al. 2019). Next, we will explain how accessibility related to spatial geography, availability of retail formats, and affordability are linked to the performance of supply chains.

14.2.2   THE LINK BETWEEN ACCESSIBILITY, AVAILABILITY, AND AFFORDABILITY

In general, food environments relate physical, socio-economic, and cultural factors to motivate or to hinder purchase, preparation, and consumption (Development Initiatives Poverty Research Ltd. 2020). Urbanization rates are growing worldwide, primarily in megacities found in developing countries (Taniguchi, Thompson, and Yamada 2016), and this growth carries not only higher congestion rates (Louf and

Barthelemy 2014), but also affects distribution operations that serve food distribution channels and food environments (Taniguchi, Thompson, and Yamada 2016; Development Initiatives Poverty Research Ltd. 2020). Food distribution becomes more difficult if the nature of the product is perishable and if the supply chain network design serves both a large variety of retail formats and direct to end consumers in a metropolitan area. In the food supply chains, delivering accurately, quickly, and efficiently becomes a competitive advantage and allows fruits, vegetables, and other fresh goods to arrive with the best quality to customers.

In this environment, geographic and demographic conditions, as well as the retail landscape and its sourcing channels, play crucial roles in helping the market (i.e., individuals and households) to obtain various products conveniently. The retailer's location is key to provide accessibility and convenience for end consumers to shop for food staples and groceries. To study food accessibility, spatial geography may be used to characterize the impact of road network, urban morphology, and densification. Spatial geography links geographical metrics and correlations to understand spatiotemporal interactions in the retail landscape for multiple formats via catchment areas and with supply chain operations by connecting food supply (i.e., availability) to the customer's demand.

Retail footprint drives urban community food systems, particularly when households live in areas with scarce access to retail chain and grocery stores, or where the income disparity is high (Taylor and Barreto 2020). Nowadays, the retailing landscape has shifted toward small retail formats like convenience stores and *nanostores* (small family-owned retailers as coined by Fransoo, Blanco, and Meiía-Argueta, 2017) that promote high turnover rates in non-perishable products, higher brand penetration, higher cash flow, and higher potential to grow the market share. However, this strategy has also brought some caveats such as: (1) a more fragmented retailing footprint, (2) higher commercial density with higher costs to serve, (3) the absence of proper equipment or backroom to store products, and (4) limited budget and risk aversion to sell produce items and other perishable, non-processed food (Fransoo, Blanco, and Mejía-Argueta 2017; Mejía-Argueta et al. 2019).

In Colombia, the advent of specialty stores (*fráver* stores) and *nanostores* that sell fruits and/or vegetables is a reality that has helped to decentralize food supply chains; but most of the time, at a higher cost than what vulnerable population segments (i.e., those living with less than US$8 per day) can afford. Thus, to guarantee the availability of produce items on the shelves of retailers worldwide, high-performance supply chains must be configured to deliver fruits and vegetables equally in the cities. To achieve this, *nanostores* should play a significant role in easing accessibility to end consumers given their unparalleled location, just a few blocks from most of the households from emerging markets. Nevertheless, decision- and policymakers need to explore how retailing connects to spatial geography to promote equity in the physical and economic accessibility for all population, while taking advantage of characteristics from distinct retail formats and connecting to their supply chains organically.

The role of supply chains is crucial to connect intermediaries, growers or producers, to end consumers via retailers. Efficient supply chains can make fruits and vegetables available at affordable prices for everyone, particularly for geographically isolated communities. High-performance supply chains help to decrease inequities in accessibility, and to increase the availability of a larger variety of food groups while satisfying tastes and preferences for low-income populations in developing countries. However, supply chain management for food and agri-business in Latin America and the Caribbean relies on traditional harvesting and production processes from growers usually located far from main cities. Along with poor infrastructure, an atomized and old fleet of carriers, and an inventory-pushed system that grants a high level of service but that is costly, the supply chain in this area becomes compromised.

14.2.3 COUPLING SUPPLY AND DEMAND FOR FRUITS AND VEGETABLES IN FOOD ENVIRONMENTS

Several authors have used geographical information systems to investigate the link associated with retail location or absence to food insecurity issues, via spatial geography, spatial clustering, and statistics (Shannon 2015; Widener 2018). Other authors (e.g., Winkenbach et al. 2018) have used geographic metrics like centrality to analyze the effect on distribution and supply chain network design for retailers. Thus, to understand better how food supply chains and food environments match, studying retailing is crucial.

Spatial analysis allows for understanding how retail location influences individual's consumption behavior and distribution channels (Roig-Tierno, Baviera-Puig, and Buitrago-Vera 2013). The composition of different retail formats in a geographic area significantly changes individuals' purchasing behavior, food choices, and consumption of certain food groups (Qin et al. 2019). Thus, catchment areas become an important leverage point to study performance of retailers to be preferred by end consumers and to investigate effects of changes in their stores and their distribution network (Widaningrum 2015).

For instance, spatial clustering shows the similarity of diverse attributes like food insecurity among neighboring units and lack of accessibility to growers or suppliers (Amarasinghe, Samad, and Anputhas 2005). On the other hand, the absence of food diversity per area might be also explored as regards to the implementation of customized policies and subsidies to directly support low-income households based on different needs. In all these analyzes, the importance of supply chain management to act as an end-to-end connector is key to make food ecosystems work properly (Development Initiatives Poverty Research Ltd. 2020), and the use of geographical information systems to address food insecurity issues gains importance. For further details, see the literature review from Caspi and colleagues (2012).

14.2.4 GAPS AND CONTRIBUTIONS

Based on this literature review, it is evident that there is a strong interdependence among retail or food environments (aggregate demand), food sourcing (supply), and nutrition (individual-based demand). Supply chains act as enablers of the food ecosystem to improve its performance to avoid food waste, combat food malnutrition, and ensure accessibility, availability, and affordability. However, there is scarce research into how these systems work in emerging markets, with particular emphasis on small, family-owned retailers or fruit and vegetable retail chains. The link between nutrition, geography, and supply chain management has been addressed only partially. In this research, we aim to connect these disciplines and explore the geographic accessibility and its effects in two cities from Sabana Centro in Colombia.

14.3 METHODOLOGY

This section introduces the steps of the process we followed to derive managerial insights, and the methods applied to this research. A detailed explanation of the data collection is presented together with the methodological framework. Also, we contemplate and justify the use of different tools as well as their relationships and expected outcomes.

14.3.1 DATA COLLECTION

Data were collected from primary and secondary sources during 2019 and 2020 (amidst the COVID-19 pandemic). On the one hand, surveys were applied to both households and *nanostore* owners to gather primary data about their operations, locations, and goods sold. On the other hand, secondary data were collected from government-based websites, databases, and public reports from recognized institutions on food insecurity, distribution channels, and statistics from Colombia.

The first data collection was carried in 2019, before the COVID-19 pandemic was announced worldwide or had even arrived in Colombia. A survey was conducted in collaboration with the workgroup of Administration and Planning for the central region of Colombia (RAP-E), which is composed of the states of Bogotá, Boyacá, Cundinamarca, Huila, Meta, and Tolima. We conducted over 500 surveys to households living in Chía and Cajicá. The survey included 85 questions about socio-economic characteristics and consumption habits of 10 food groups framed in the regional basic food basket. The goal was to identify patterns of the households' consumption behavior, perceived food environments, and preferences of fresh food (fruits and vegetables) depending on cultural and consumption habits.

The secondary sources included socio-economic and geographical data from the National Statistics Institute of Colombia (DANE), and RAP-E. DANE supplied geo-localized data about socio-economic levels, while the workgroup of RAP-E helped to design the items and features of the survey (e.g., type, sample size, randomized sampling methods, and localizing geographically collected data) (DANE 2021).

Additionally, we geo-referenced all retailers' locations in the area under study using Google Maps™. A total of 186 *nanostores*, 78 *fruvers*, and 99 supermarkets from Chía and Cajicá were considered in our investigation. For supermarkets, we classified them into convenience stores (i.e., usually smaller than 500 m²), traditional stores (i.e., bigger than 500 m²), and large supermarkets (usually called 'hypermarkets') (Farley et al. 2009) to analyze their impact according to their characteristics.

14.3.2 CONCEPTUAL FRAMEWORK

We structured the collected data analyzing the retail formats from the geographical area under analysis. We observed differences in the geographical features, demographic and socio-economic data, as well as in retail composition. We describe these features in the next three subsections.

14.3.2.1 Geographical Attributes

To identify the initial spatial patterns of retailers' distribution within both municipalities, we adopted the urban landscape's fragmentation metrics made by Angel, Parent, and Civco (2012) and Gavrilidis and colleagues (2016). These metrics are relevant to connect spatial geography with logistics. We employed a *k-median* clustering to divide the selected regions into nine clusters, which allowed us to classify the territory based on the geographical position by using fragmentation metrics. Figure 14.2 illustrates this classification by showing a subdivision of both municipalities into "*veredas*" (such as Canelon, Hato Grande, Fagua, Tiquiza, among others). A *vereda* represents several small geographical divisions for administrative purposes. Considering these clusters and according to the urban land use purpose (e.g., residential, business, industrial area), five critical areas are defined: (1) downtown, (2) urban extension, (3) urban areas under geographical barriers for the municipality, (4) developing areas, and (5) urban bridges.

The urban built-up area refers to the downtowns in both municipalities, where there is higher population density, and where the geographical center is found (see clusters 6 and 7 in Figure 14.2). The urban extension is around downtown. In this area, the population density is high but less than in downtown (clusters 3 and 4). The areas bordering the city limits are called urban barriers as they impose administrative constraints on the municipality's growth (see clusters 0 and 1). The peri-urban areas— areas between urban and rural areas—are named developing areas. They are the territory assigned to promote municipality's growth for residential urban functions (see clusters 2 and 8) (Gavrilidis et al. 2016). Finally, urban bridges are geographic areas that are not formally defined in the urban landscape. They refer to intersections between two or more administrative divisions such as municipalities, cities, *veredas*, etc. They are transitional areas between these divisions/municipalities (see cluster 5).

14.3.2.2 Demographic and Socio-economic Characteristics

For our goal, two features are used to characterize each cluster and geographical area: socio-economic level and population density. The former refers to a general classification used in Colombia to mark differences among several neighborhoods depending on the median income level of households. This classification is made by the office of land use and property registry with the purpose of determining different rates of public utilities based on the conditions and physical characteristics of the dwelling and its surroundings. Thus, this is an indicator of a household's purchasing parity (DANE 2015; Guzman, Arellana, and Cantillo-Garcia 2020). The socio-economic levels in Colombia are between 1 (the lowest class—extreme poverty) to 6 (the wealthiest class—upper level). For the area under study, the classification is

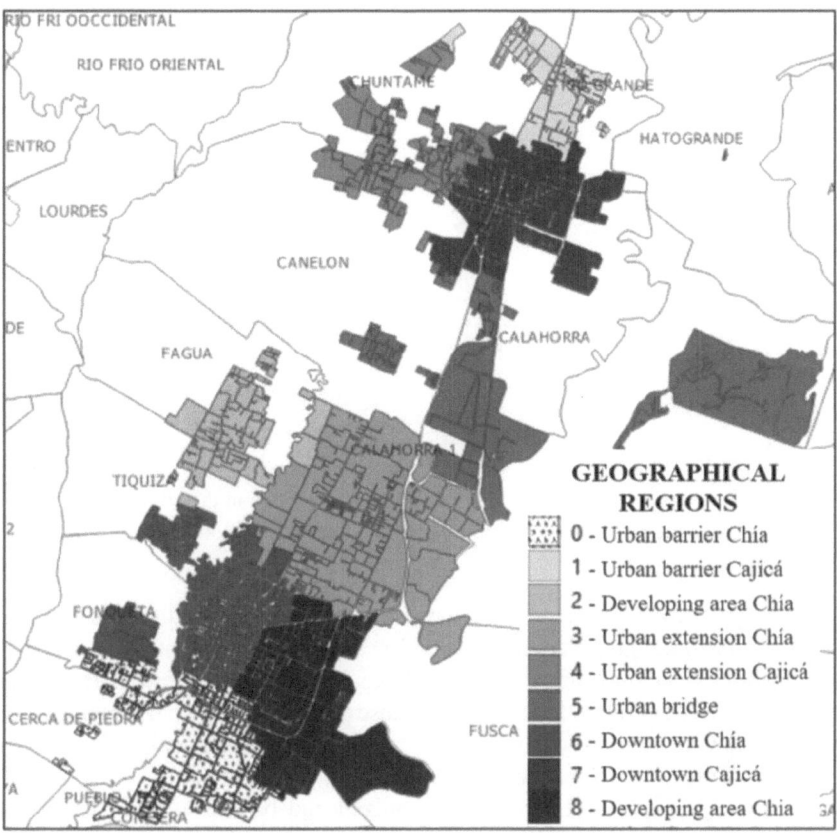

FIGURE 14.2 Representation of geographic classification of the selected regions showing divisions by *veredas*

shown in Figure 14.3. The predominant strata are 2 and 3 (i.e., low and mid–low classes—vulnerable population segments) in both municipalities.

The population density is calculated by DANE (2021). Figure 14.3 displays the population density per neighborhood in both municipalities. It is clear that in downtown areas, there is a much higher density in comparison with the peripheral areas.

14.3.2.3 Retail Landscape

Considering the geographic and demographic overview, we looked at three types of retail formats that are found in the case region landscape: *nanostores*, supermarkets, and *fruvers*. These stores are the main formal sellers of fruits and vegetables in both municipalities.

A *nanostore* is a small unit usually owned and run by a household which sells an assortment of food products and other basic goods. Depending on the size, these small businesses sell fresh and packaged foods, and a variety of miscellaneous items such as stationery, phone cards, and cigarettes, to name a few. Owners of *nanostores* normally sell

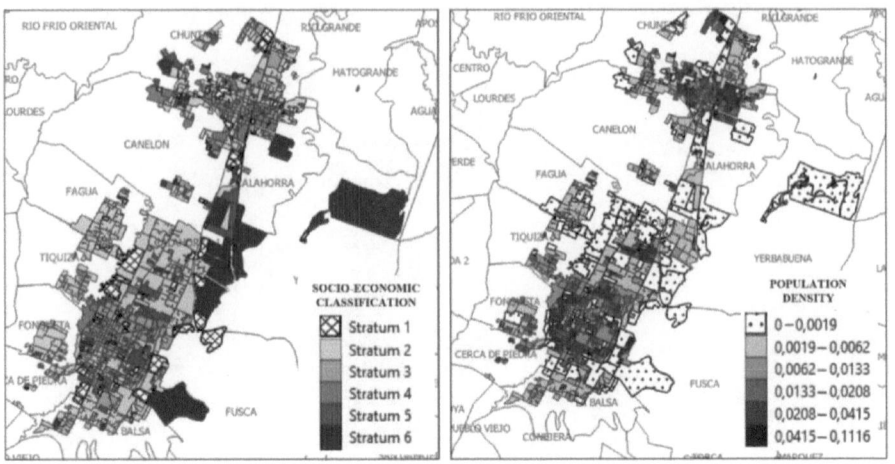

FIGURE 14.3 Representation of demographic characteristics in the study area: *(left)* socio-economic classification according to stratums; and *(right)* population density

produce from "behind the counter," although some have a handful of aisles and shelves. A few *nanostores* use information technologies, but most of them only accept cash. Due to the limited space, a *nanostore* is supplied two or three times per week by consumer-packaged goods (CPG) manufacturers. In the case of fresh products, shopkeepers travel to buy them a few times per week, using the facilities of major wholesalers like the Central Market (i.e., Corabastos) (Fransoo, Blanco, and Mejía-Argueta 2017).

In contrast, supermarkets are well organized and belong to retail chains. They usually have an aisle and shelf layout, provide carts and baskets to ease purchases from shoppers, and accept several payment methods (e.g., cash, electronic payments, vouchers). "*Autoservicios*" is the smallest supermarket type, and it can be seen as an enhanced, automated *nanostore* given its small size. A recently introduced format is the hard-discount store, which shows a similar layout and size to *autoservicios,* but they offer lower prices as they take advantage of economies of scale in purchasing, sourcing, and distribution in large quantities.

Frúver stands for "frutas y verduras" in Spanish (i.e., fruits and vegetables). This type of business started about 20 years ago as a family-owned business that only sells fresh fruits, vegetables, and sometimes, meat, fish, and dairy products. The original brand "Surtifruver" has successfully expanded its business model throughout Colombia. Store sizes vary, but in general they resemble *autoservicio's* store sizes (50–200 m²). *Fruvers* purchase directly from farmers in the countryside, unlike *autoservicios*, hard-discount stores, and *nanostores*. *Fruvers* regularly accept similar payment methods to supermarkets.

14.3.3 DATA MODELING AND TOOLS FOR ANALYSIS

To model the collected data, we used the open source, geographical information system called Qgis 3.16.0. Using this software, we analyzed the spatial distribution and other geographical patterns using three types of tools: (1) catchment areas (i.e., buffer rings), (2) spatial clustering, and (3) Voronoi diagrams.

First, we defined the catchment areas through the multi-ring buffers that were created for diverse retail formats. These rings helped us to discover geographical patterns in both municipalities. Then, the relation between geographical and demographic features is performed for selected areas and street blocks by using hierarchical clustering tools like the Attribute Based Clustering complement from the SciPy library of Python. As described in Section 14.3.2.2, the target variables to generate the clustering were population density and socio-economic levels. The clusters were defined to compare those with the presence of certain retail formats with those with an absence. However, information about the real catchment area, and the density of each retail channel was still needed to counteract with these data. Lastly, Voronoi diagrams were employed for all retail formats. The granularity found in the polygons provided an adequate approach to understand the real influence area of distinct retailers around the selected regions. A detailed explanation for each tool is given below.

14.3.3.1   Catchment Areas and Buffer Rings

A catchment area is the area around a business, facility, or location that covers, or "catches" from where its customers, households, or individuals come from (Cambridge Dictionary Web 2021). Buffer ring analysis is used as the first approach to analyze the influence of location and distance or time on coverage. This analysis is a simple and common method used to find the influence of an entity on a defined area of coverage at a given radius. This influence zone may be estimated using two basic methods: the Euclidian and the Geodesic. The former is the most used, which calculates distances between geographical points in a bidimensional cartesian plane. In the latter method, distances are calculated considering the earth curvature (Oliver, Schuurman, and Hall 2007). For our case, we used Euclidian distances.

Buffer rings define catchment areas of the potential customers served by each retail channel in the selected region. In our case, smaller retailers, such as *fruvers* and *nanostores*, should have a catchment area of up to 300 meters in radius around the store's location. On the other hand, supermarkets can serve up to 2 km around its location (Aránguiz, Mejía, and Mejía-Argueta 2018). However, due to the size of the selected regions, we computed buffers of 1 kilometer of radius around large stores such as hard-discount stores and large supermarkets, while convenience stores or *autoservicios* used the same radius as small retailers. These buffer rings can be later intersected with the map and other metadata of the region.

The buffer ring method has a few limitations. For example, the buffer does not consider geographic restrictions such as the presence of topographic boundaries, rivers, and competitors, etc.. Further, the ring may not accurately represent the spatial coverage or influenced area (Landex and Hansen 2006).

14.3.3.2   Hierarchical Clustering

Clustering consists of placing observations or entries into groups (i.e., clusters) of similar characteristics. Clustering algorithms segment a dataset into groups with maximal similarity within clusters, and minimal similarity among clusters (Larose and

Larose 2015). In the case of hierarchical clustering, a dendrogram (tree-shaped structure) is proposed through split (starting with one cluster and dividing into groups) or combination (starting with multiple single clusters and grouping) methods of clustering. The determination of distance among clusters is critical for this algorithm. Three methods are commonly used: (1) the single linkage, based on the shortest distance or nearest neighbor method; (2) the complete linkage, which considers the largest distance among clusters; and finally, (3) the average linkage, which considers the average distance of records from one cluster to another (Larose and Larose 2015).

We used hierarchical clustering because it finds the adequate number of clusters according to the input variables that were demographic features. Further, hierarchical models are more stable with the increase of outliers, and some cases were found where hierarchical clustering delivered better results when incidents of overlapping data are presented (Milligan 1980).

14.3.3.3 Voronoi Diagrams

A Voronoi diagram is a partition of a region into cells and partitions (i.e., Voronoi cells), given a finite set of n geographical points $\{1, ..., i, j, ..., n\}$, and a definition of distance (rectilinear, Euclidean, road). Without loss of generality, each geographical point i is defined in the two-dimensional space and its coordinates are x_i, y_i. Let S be a region and let d_{Si} be the set of all distances between all points in S and a geographical point i. Formally, a Voronoi cell (VCi) associated to the geographical point i is defined as (Melo, Frank, and Brantingham 2017):

$$VC_i = \{S | d_{Si} \leq d_{Sj}, i \neq j\}$$

Therefore, in all points in a cell VC_i, the distance to point i is less than or equal to the distance to another point j. Then, a Voronoi diagram (VD) is defined as the set of all cells VC_i.

$$VD = \{VC_i, \forall i\}$$

The size of the cells is related to the density of the region under study: the smaller the cells, the higher the density (in points) of the region. In our case, this is the density of food retailing (i.e., food environment) in the selected regions and as we mentioned before, this shows the real influence of each retailer in both municipalities. This analysis, joined with the catchment areas and hierarchical clustering analysis, evaluated the initial found patterns and helped to derive final conclusions.

14.4   RESULTS AND ANALYSIS

14.4.1   PRELIMINARY DISTRIBUTION PATTERNS

Figure 14.4 shows the buffer rings which determine the catchment areas for each retailer type. As we mentioned in the methodology, *nanostores, fruvers,* and small supermarkets (i.e., *autoservicios,* convenience stores) have an influence area smaller than the large retailers. Considering this assumption, we saw that *fruvers* are mainly located in downtown areas and urban extensions with immediate access to main roads. There is only a handful of *fruvers* in developing areas. Thus, Figure 14.4 illustrates a first pattern, in which *fruvers* tend to be present in areas with a high population density.

For *nanostores*, there is not a specific pattern in their location: their presence is observed in all clusters except for the urban bridge and part of a developing area (see also Figure 14.2). Curiously, these areas correspond to wealthier areas (according to Figure 14.3). Therefore, we can conclude that *nanostores* locate primarily in low–medium income areas.

Finally, buffer rings of supermarkets show the considerable influence of large retailers in certain clusters. The more significant part of convenience stores (i.e., smaller supermarkets) is located in downtowns (the smaller buffer rings in Figure 14.4); while other supermarkets cover almost all geographical clusters. However, there are no supermarkets in developing areas (in the western part of cluster 2—see Figure 14.2). Unsurprisingly, big supermarkets are the only retail format that has a presence in the urban bridge (one of the wealthiest areas).

14.4.2 SOCIO-ECONOMIC CLUSTERING ANALYSIS

As a result of the clustering through the hierarchical method, five clusters were generated, as presented in Figure 14.5. This method weighed the input variables (socio-economic level and population density) using the complete linkage method and calculating the distance through the Euclidean metric.

Three of the five clusters corresponded to stratum 2 (see also Figure 14.3), which is highly dominant in the selected region (see clusters 1, 2, and 4 in Figure 14.5). Clusters 1 and 2 had low presence of all retailers. Interestingly, both clusters were

FIGURE 14.4 Buffer rings to determine catchment areas for each retailer type: *(left)* *nanostores*, *(middle) fruvers*, and *(right)* supermarkets

FIGURE 14.5   Clusters grouped by the hierarchical method

composed of small, isolated blocks in peripheral regions. However, cluster 2 presents fewer retailers (it mostly shows presence of *nanostores*), while cluster 1 did not have the direct presence of any type of retailer. Cluster 1 is by far the most densely populated, where the predominant stratum is 2. This cluster is located in the development areas of both municipalities. Although these regions have no identified retail channels, nearby clusters, usually found as cluster 2 or 4, have enough retailers to serve these regions.

Cluster 3 is predominantly formed by blocks with neighborhoods of strata 4 and 6 (mid–high and high classes). In fact, this was the only cluster that had lower participation of *nanostores* in comparison with other retail formats. There, it was possible to only count three *nanostores* and three *fruvers*, whilst 10 supermarkets were found to cover the cluster. Primarily, the density of supermarkets is over three times the density of the *nanostores* and *fruvers*, as Table 14.1 shows.

Cluster 4 is found only in neighborhoods of stratum 2 (low-income level), which was detected around the downtown areas and in urban extensions composed of small blocks. Finally, the blocks around the downtown areas were strategically clustered together. There, the predominant stratum was 3 (mid–low income), which has presence of the three retail formats. Hence, the urban planning and nearby location to higher density areas might favor the location of all types of retailers. Importantly, the presence of *nanostores* was twice higher than any other retail format, adding a total of 103 stores compared to 47 *fruvers* and 55 supermarkets.

TABLE 14.1
Density of Retailers in Each Hierarchical Cluster Classification

Hierarchical Cluster Classification	Density of *Nanostores* (# of *nanostores*/km²)	Density of *Fruvers* (# *fruvers*/km²)	Density of Supermarkets (# supermarkets/km²)
1	0	0	0
2	47.06	13.72	11.77
3	0.31	0.31	1.03
4	3.86	1.31	1.83
5	19.94	9.10	10.65

There is a significant larger presence of *nanostores* in all clusters except for cluster 3 where supermarkets dominate. In cluster 3, which is composed of strata 4 and 6, supermarkets are the most common retail choice due to the easy access to private vehicles and the higher purchasing parity of wealthier households.

In cluster 5, a *nanostore* can serve from 760 to roughly 1,085 individuals. Considering that the average size of a household is around four people, this means that in downtown, *nanostores* can serve from 190 to 200 families on average. In the same region, *fruvers* should reach from 450 to 560, and supermarkets reach to only around 400 households.

14.4.3 DEMAND AND SUPPLY ANALYSIS

In the surveys collected in collaboration with RAP-E, we included strata 1 to 4, which are the most likely to suffer from food insecurity. The weekly consumption of fresh food (i.e., fruits and vegetables under the survey's classification) per household is on average between 1 and 2 kg. Stratum 1 shows the highest average of 1.88 kg of consumption per family, and stratum 3 shows the lowest value of 1.20 kg per family. Figures 14.6 shows Voronoi diagrams for the presence of *fruvers*, supermarkets, and *nanostores* in Chía and Cajicá. Each of the diagrams relates to one retail format, and the intensity of color represents the density of that retailer in the distinct geographical areas using a logarithmic scale.

For *fruvers,* Figure 14.6 shows that the highest density occurs in the downtown areas, which we also observed in the catchment areas analysis; while the lowest density occurs in the peri-urban areas (i.e., developing areas and urban barriers) in western Cajicá. In addition, the highest density of *fruvers* corresponds to areas of highest population density (see also Table 14.1 and Figure 14.3).

In the case of supermarkets, Figure 14.6 presents a different behavior in the two regions. In Cajicá, this format is predominantly located in downtown. On the other hand, in Chía, even though there is a high density of supermarkets closer to the downtown, it is possible to find a high presence of supermarkets around the urban bridge, and closer to wealthier areas (higher strata), where both *fruvers* and *nanostores* have no presence. This conclusion confirms what we detected in the preliminary pattern

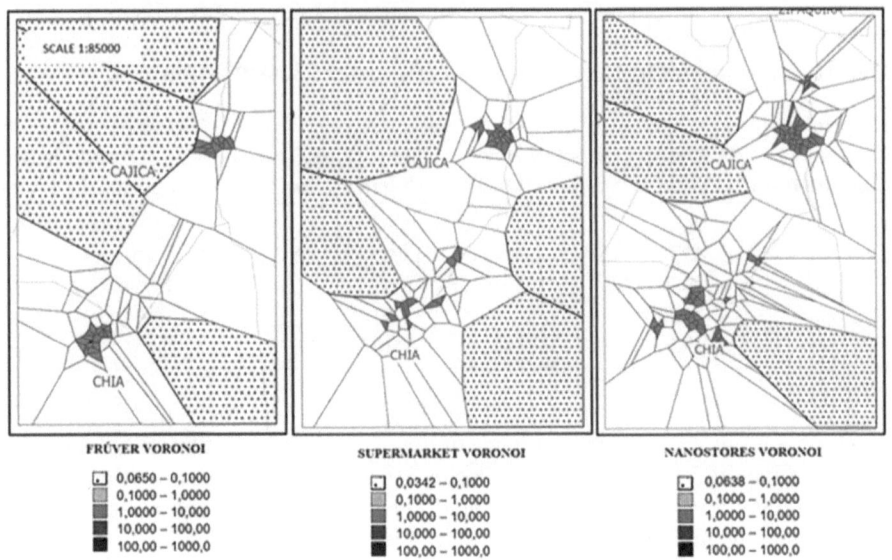

FIGURE 14.6   Voronoi diagram per retail format in Chía and Cajicá: *(left) fruvers, (middle)* supermarkets, and *(right)* nanostores

analysis from the buffer rings analysis. There is a low granularity in geographical barriers, developing areas, and urban extensions, which can be understood as a low density of retailers in those regions.

Finally, Figure 14.6 indicates the presence of *nanostores* in both municipalities, with a pattern that is remarkably similar to the one found with *fruvers* around down-town; however, with a higher density and granularity than *fruvers*. Another highlight of this analysis is that *nanostores* present high-density levels farther away from the town centers (i.e., in fringe areas like barrier areas and urban extensions), where the density is noticeably higher than other retailers. Unsurprisingly, these areas are dominated by socio-economic stratum 2 (low-income population).

14.5   CONCLUSIONS

The proposed approach allowed the understanding of the distribution of the main retail channels found in the first stage of the study and the analysis of relationships among socio-economic factors (such as poverty and population density) to the presence of the retailers.

Considering the location, *fruvers* and supermarkets were mostly driven to address higher density areas or (mostly supermarkets) high-income areas. *Nanostores,* mean-while, do not present a clear geographical growth structure, which usually responds to a need of the establishment's owner, with few or no study before implementation. This might also show that they may be capturing the demand of households living in fringe areas, where other retailers are not present. However, their assortment might be limited and the accessibility of fruits and vegetables for those communities might be limited and non-affordable.

For example, it was possible to notice that the *nanostores'* presence in peripheral low-income areas makes this format an important retailer for accessibility, availability, and affordability of fresh food in those under-served areas. Considering that it may be difficult to predict where a new *nanostore* will appear, its proliferation and behavior in time must be studied so that fruits and vegetables are made available for the most vulnerable population segments. In addition, if it is desired that *nanostores* have a clear role in the supply of fresh food for the most vulnerable regions (peripheral, with low resources), actions must be taken, so that the owners of *nanostores* find greater motivation to improve the assortment and quality of the fresh food offered, without hindering their basic operations in the management of their establishment.

Further studies in this direction should add more dynamic variables, such as capacity of migration among diverse retail formats, retention of customers or self-supply of the consumers, and the appearance of other types of retailers in the studied regions, such as local producers or mobile markets. Similarly, price and quantities have to be studied carefully per retail format, to derive intervention schemes to improve availability and affordability based on different food groups and individual tastes and preferences. Finally, it is crucial to investigate how to configure more effective supply chains to link the under-served food environments to farmers and other sourcing methods with a smaller number of intermediaries.

REFERENCES

Amarasinghe, Upali, M. Samad, and M. Anputhas. 2005. "Spatial clustering of rural poverty and food insecurity in Sri Lanka." *Food Policy,* 30 (October): 493–509.

Angel, Shlomo, Jason Parent, and Daniel L. Civco. 2012. "The fragmentation of urban landscapes: Global evidence of a key attribute of the spatial structure of cities, 1990–2000." *Environment and Urbanization,* 24 (1): 249–283.

Anthem, Paul. 2020. "Risk of Hunger Pandemic as Coronavirus Set to Almost Double Acute Hunger by End of 2020." United Nations World Food Programme. www.wfp.org/stories/risk-hunger-pandemic-coronavirus-set-almost-double-acute-hunger-end-2020.

Aránguiz, Raúl, Gonzalo Mejía, and Christopher Mejía-Argueta. 2018. "Encuesta a Tiendas de Barrio y Estrategia de Ubicación de Mercados Móviles En Valparaíso." Master Thesis. Pontifical Catholic University of Valparaíso, Chile.

Bauerová, Radka. 2018. "Consumers' decision-making in online grocery shopping: The impact of services offered and delivery conditions." *Acta Universitatis Agriculturae et Silviculturae Mendelianae Brunensis,* 66 (5): 1239–1247.

Cambridge Dictionary Web. 2021. "Catchment Area Meaning." https://dictionary.cambridge.org/es/diccionario/ingles/catchment-area (accessed January 24, 2021).

Caspi, Caitlin E., Glorian Sorensen, S. V. Subramanian, et al. 2012. "The local food environment and diet: A systematic review." *Health Place,* 18 (5): 1172–1187.

DANE (National Statistics Institute of Colombia). 2021. "Geoportal DANE – Descarga Del Marco Geoestadistico Nacional (MGN)." https://geoportal.dane.gov.co/servicios/descarga-y-metadatos/descarga-mgn-marco-geoestadistico-nacional/ (accessed January 24, 2021).

DANE (National Statistics Institute of Colombia). 2015. "Metodología de Estratificación Socioeconómica Urbana Para Servicios Públicos Domiciliarios." www.dane.gov.co/files/geoestadistica/estratificacion/EnfoqueConceptual.pdf.

Dannefer, Rachel, Donya A. Williams, Sabrina Baronberg, et al. 2012. "Healthy Bodegas: Increasing and promoting healthy foods at corner stores in New York City." *American Journal of Public Health,* 102 (10): e27–31.

Development Initiatives Poverty Research Ltd. 2018. "2018 Global Nutrition Report." https://globalnutritionreport.org/reports/global-nutrition-report-2018/.

Development Initiatives Poverty Research Ltd. 2020. "2020 Global Nutrition Report." https://globalnutritionreport.org/reports/2020-global-nutrition-report/.

FAO, FIDA, OPS, WFP, and UNICEF. 2020. "Panorama de La Seguridad Alimentaria y Nutrición En América Latina y El Caribe 2020." www.fao.org/documents/card/en/c/cb2242es/.

FAO, IFAD, UNICEF, WFP, and WHO. 2020. *The State of Food Security and Nutrition in the World 2020. Transforming food systems for affordable healthy diets.* Rome: Food and Agriculture Organization of the United Nations.

Farley, Thomas A., Janet Rice, J. Nicholas Bodor, et al. 2009. "Measuring the food environment: Shelf space of fruits, vegetables, and snack foods in stores." *Journal of Urban Health,* 86 (5): 672–682.

Fransoo, Jan C., E. E. Blanco, and C. Mejía-Argueta. 2017. *Reaching 50 Million Nanostores: Retail distribution in emerging megacities.* Scotts Valley, CA: CreateSpace.

Gavrilidis, Athanasios Alexandru, Cristiana Maria Ciocănea, Mihai Răzvan Niță, et al. 2016. "Urban landscape quality index – Planning tool for evaluating urban landscapes and improving the quality of life." *Procedia Environmental Sciences,* 32: 155–167.

Guzman, Luis, Julián Arellana, and Victor Cantillo-Garcia. 2020. "El Estrato Socioeconómico Como Variable Sustituta del Ingreso en La Investigación en Transporte. Evaluación Para Bogotá, Medellín, Cali y Barranquilla." www.researchgate.net/publication/343426887_El_estrato_socioeconomico_como_variable_sustituta_del_ingreso_en_la_investigacion_en_transporte_Evaluacion_para_Bogota_Medellin_Cali_y_Barranquilla/link/5f29860ca6fdcccc43aa5e19/download.

Landex, Alex and Stephen Hansen. 2006. "Examining the Potential Travellers in Catchment Areas for Public Transport." Proceedings of ESRI International User Conference. http://proceedings.esri.com/library/userconf/proc06/papers/papers/pap_1391.pdf.

Larose, Daniel T. and Chantel D. Larose. 2015. *Data Mining and Predictive Analytics.* 2nd edition. Hoboken, NJ: Wiley.

Louf, Rémi and Marc Barthelemy. 2014. "How congestion shapes cities: From mobility patterns to scaling." *Scientific Reports,* 4.

Mejía-Argueta, C., V. Benítez-Pérez, S. Salinas-Benítez, et al. 2019. "Nanostores, a Force to Reckon With to Fight Malnutrition." European Supply Chain Forum. www.linkedin.com/pulse/nanostores-force-reckon-fight-malnutrition-escf-professors/?fbclid=IwAR2t_KpdYvMfbaId_hlZ3QLQsSpO6CceKX2eXHLVhaQgui-ZCLFsChD6lgw.

Melo, Silas Nogueira de, Richard Frank, and Patricia Brantingham. 2017. "Voronoi diagrams and spatial analysis of crime." *The Professional Geographer,* 69 (4): 579–590.

Milligan, Glenn W. 1980. "An examination of the effect of six types of error perturbation on fifteen clustering algorithms." *Psychometrika,* 45 (3): 325–342.

Oliver, Lisa N., Nadine Schuurman, and Alexander W. Hall. 2007. "Comparing circular and network buffers to examine the influence of land use on walking for leisure and errands." *International Journal of Health Geographics,* 6: 1–11.

Paluta, Lauren, Michelle L. Kaiser, Sarah Huber-Krum, et al. 2019. "Evaluating the impact of a healthy corner store initiative on food access domains." *Evaluation and Program Planning,* 73: 24–32.

Qin, Yong Jun, Nathanaël Pingault, F. Ricci, et al. 2019. "Improved food environments for healthy diets and enhanced nutrition." *Journal of Integrative Agriculture*, 18 (7): 1652–1654.

RAP-E (The Administrative and Special Planning Region). 2021. "Plan de Abastecimiento Alimentario de La Región Central." https://regioncentralrape.gov.co/plan-de-abastecimiento-de-la-region-central/ (accessed January 24, 2021).

Roig-Tierno, Norat, Amparo Baviera-Puig, and Juan Buitrago-Vera. 2013. "Business opportunities analysis using GIS: The retail distribution sector." *Global Business Perspectives,* 1 (3): 226–238.

Sabana Centro Cómo Vamos. 2018. "Informe de Calidad de Vida 2018." http://sabanacentrocomovamos.org/home/wp-content/uploads/2019/11/4to-Informe-de-Calidad-de-Vida-de-Sabana-Centro_2018.pdf.

Shannon, Jerry. 2015. "Rethinking food deserts using mixed methods GIS." *CityScape,* 17 (1): 85–96.

Taniguchi, Eiichi, Russell G. Thompson, and Tadashi Yamada. 2016. "New opportunities and challenges for city logistics." *Transportation Research Procedia,* 12: 5–13.

Taylor, Jamal and Luiz Paulo Silva Barreto. 2020. "Closing the food access gap in American underserved communities." Master Thesis. MIT Center for Transportation and Logistics, U.S. https://dspace.mit.edu/handle/1721.1/128251.

Widaningrum, Dyah Lestari. 2015. "A GIS-based approach for catchment area analysis of convenience store." *Procedia Computer Science,* 72: 511–518.

Widener, Michael J. 2018. "Spatial access to food: Retiring the food desert metaphor." *Physiology and Behavior,* 193: 257–260.

Winkenbach, Matthias, Daniel Merchán, Milena Janjevic, et al. 2018. "City Logistics Policy Toolkit: A Study of Three Latin American Cities." https://documents.worldbank.org/pt/publication/documents-reports/documentdetail/413421568727734853/city-logistics-policy-toolkit-a-study-of-three-latin-american-cities.

World Health Organization. 2020. "Malnutrition." www.who.int/news-room/fact-sheets/detail/malnutrition.

15 Analysis of Internet of Things Implementations Using Agent-based Modeling
Two Case Studies

Mohammed Basingab, Khalid Nagadi,
Atif Shahzad, and Ghada Elnaggar

15.1 INTRODUCTION

The Internet of Things (IoT) has emerged as an entirely new domain—although existing previously—relying entirely on the connectivity among physical objects of every kind with abilities to sense, communicate, intelligently analyze, and respond as part of a larger system. Classical fields with a wide-spread adoption have found novel applications and unprecedented ways of implementing IoT with the growing use of smart technologies. Broadly speaking, the application areas can be divided into consumer market-related applications, commercial applications, industrial applications, and applications in the infrastructure (Perera, Liu, and Jayawardena 2015).

This chapter focuses on the application area of maintenance in the medical field. IoT and agent-based modeling (ABM) are used in two case studies presented in this study. ABM is a widely used modeling technique in the domain of computational modeling for an effective implementation using simulation. This involves the use of autonomous agents interacting with each other for the exchange of continuously generated data and the effect of the various interactions on the system. ABM is ideally suited for modeling IoT applications due to similar nature such as: autonomous actions of the agents and the things; communication among things corresponding to interaction among agents; and effects of the dynamic changes in the system observed through sensing capabilities of the things analogous to intelligent observatory and assessment capabilities of the agents. In addition, simultaneous interaction of agents makes the adaptive process in complex systems, such as IoT, more scalable. Agents, emergence and complexity form the basis of ABM. Agents are usually individual entities; however, they can be viewed as a collection or a cohort. Emergence is certain characteristics uniquely observed within the agents only when these agents are capable of interacting with each other in a system and play the desired role expected

DOI: 10.1201/9781003137993-16

by the system. By complexity, we mean the multi-faceted interaction among agents that follows certain rules of interaction at local levels.

The use of ABM-based simulation is widely accepted in various domains. As the interaction among agents becomes more complex due to an increased number of agents, simulation using ABM becomes a more viable solution to analyze the system behavior and to identify the characteristics of the system in various possible situations. For example, in the case of maintenance, the failure rate of the components can be varied in accordance with the realistic situations that occur as the system interactions take place. Reducing the failure rate is usually attributed to higher costs through the installation of smart devices providing observation for a better predictive maintenance plan generation. Now this higher cost is later justified by a better understanding of the system behavior in the long run. This subsequently reduces the overall associated cost as the system is expected to have a robust operation.

Simulation can be effectively used to build various scenarios in situations to develop a comprehensive understanding of the system response. In this study, AnyLogic® is used for the simulation of the systems described in both case studies. AnyLogic is inherently suited for simulation with ABM implementation in contrast to the traditional process modeling approach.

15.2 RELATED WORK

Computational advances made wide use of ABMs across many application areas that support decision making (Macal and North 2009). Examples of these applications are supply chains, predicting the spread of epidemics, mitigating the threat of bio-warfare, and understanding the factors that may be responsible for the fall of ancient civilizations.

Grimm and colleagues (2006) explained a framework of both individual-based models (IBMs) and ABMs in ecology. This framework consists of three blocks, namely: overview, design concepts, and details (ODD). ODD can be used to begin constructing a detailed individual- or agent-based model. Hinkelmann and colleagues (2010) modified the ODD framework to permit characterizing an ABM as a dynamic system. Plenty of applications of ABM can be found in the literature.

Bruch and Atwell (2015) inferred procedures to apply ABM on empirical research to conceptualize, develop, and evaluate empirically grounded ABMs. They discussed multiple modes in which ABM can be applied in simple or large-scale real systems. North and colleagues (2010) developed a consumer market model with multiple levels using ABM. It was applied by Procter & Gamble, and the results showed the effectiveness for decision making and cost savings. A review of some ABMs used in financial markets can be found in Samanidou and colleagues (2006). Showing similarities with scaling laws for other systems with many interacting units, they proved that an exploration of financial markets as multi-agent systems appeared to be a natural consequence.

In the field of biology, An and colleagues (2009) reviewed the ABM scheme and explained some uses of its applications in the field of acute inflammation and wound healing. They mentioned that due to the existence of multiple levels of organization of biomedical research that require developments in transitional methodologies,

ABM is vital. ABM is a non-inductive modeling method that focuses on the inter-action between individual components. To reinforce comprehension of immunology and disease pathology, Bauer and colleagues (2009) developed some ABMs of host-pathogen. Rockett and colleagues (2020) studied the genome sequencing of SARS-CoV-2 in a subpopulation of infected patients during the first 10 weeks of COVID-19 containment in Australia, and compare findings from genomic surveillance with predictions of computational ABM. Using Australian census data, the ABM generates over 24 million software agents representing the population of Australia, to simulate spreading from specific infection sources, using contact rates of individuals within different social contexts. Silva and colleagues (2020) propose the COVID-ABS, a new SEIR (Susceptible–Exposed–Infected–Recovered) ABM that aims to simulate the pandemic dynamics using a group of agents imitating different sectors. A range of scenarios for social distancing interventions were studied, with varying epidemio-logical and economic effects. Although scenarios with lockdown present the lowest number of deaths (and highest impact on the economy), scenarios combining the use of face masks and partial isolation may be more realistic for implementation in terms of social cooperation. The model can be considered as a benefit tool to help author-ities to plan their actions against the COVID-19 epidemic. Chapizanis and colleagues (2020) simulated human movement and interaction behavior, using ABM. A city scale ABM was developed for urban Thessaloniki, Greece that feeds into population-based exposure assessment without imposing prior bias, basing its estimations onto emer-ging properties of the behavior of the computerized autonomous decision makers (agents) that compose the city-system.

The IoT is a paradigm where everyday objects can be equipped with identifying, sensing, networking, and processing capabilities that will allow them to communicate with one another and with other devices and services over the internet to accomplish an objective (Whitmore, Agarwal, and Xu 2015). IoT is a platform where everyday devices become smarter: every day, processing becomes intelligent, and every day, communication becomes informative (Ray 2018). ABM represents a proper envir-onment to tackle the potentials of IoT (Savaglio et al. 2020). IoT is a revolutionary concept, within cyber-physical systems, that is rich in potential as well as in multi-faceted requirements and development issues. To properly address the concept, and to fully support IoT systems development, agent-based computing represents suit-able and effective modeling, programming, and simulation paradigms. As a matter of fact, agent metaphors, concepts, techniques, methods, and tools have been widely exploited to develop IoT systems (Savaglio et al. 2018). Recent applications of integrated ABM and IoT can be found in Dzaferagic and colleagues (2019), who investigated the suitability of ABM to model the communication aspects of a road traffic management system as an example of an IoT network. The results showed the necessity of coordination between multiple decision makers in order to achieve quality reporting and improved system usage.

15.3   CASE STUDY 1

A retail pharmacy facility was chosen for our research. The scope of the case study is limited to 20 pharmacies and one warehouse located in Saudi Arabia. An ABM was

TABLE 15.1
Input Parameters for the ABM

#	Parameter	Value	Unit
1	# of trucks	6	Truck
2	Distance between locations	GIS	GIS
3	Refrigerator repair time	Triangular (3, 3.5, 4)	Hour
4	# of facility warehouse	1	Warehouse
5	# of pharmacies	20	Store
6	Speed of the trucks	70	Km/h
7	Response time	Uniform (7, 9)	Hour
8	Refrigerator failure rate	Triangular (2, 3, 4)	Ref./week
9	Delay time	Uniform (1.5, 2)	Hour

TABLE 15.2
Sample of the Locations of the Pharmacies

#	Longitude	Latitude
1	39.2615459	21.5175112
2	39.2175118	21.4559082
3	21.53671	39.1934
4	21.49567	39.17784
5	21.47418	39.21247
Warehouse	21.47376	39.23901

built, using AnyLogic software, to simulate the failure behaviors of the refrigerators in the facility. Currently, the retail pharmacy facility has an approximate annual cost of $167,000 due to refrigerators failures.

The ABM contained 4 agents: pharmacy, trucks, orders, and warehouse. The input parameters are shown in Table 15.1.

All the data was collected from the facility specialists. Table 15.2 shows a sample of the exact physical locations (longitude and latitude) of the 20 pharmacy branches, including the one warehouse.

15.3.1 SIMULATION MODEL

The ABM was built using AnyLogic software. First, all the physical locations of the pharmacies were set in the model using GIS capabilities. Then, the process of ordering the maintenance team between the warehouse and the pharmacies was developed. Finally, four different scenarios of applying different levels of IoT were analyzed. ABM simulation results of the base model are summarized in Table 15.3.

TABLE 15.3
Summarized Simulation Results of the Base Model

Measure	Result	Unit
# of failed ref.	1,876	# of ref.
Out of service time	912	# of days
Total cost	167,390	Saudi Riyal/year

TABLE 15.4
A Comparison of the ABM Scenarios

Scenario	Failure Rate Reduction	Out of Service Cost (Saudi Riyal/year)	Total Cost (Saudi Riyal/year)
1	85%	1,982	23,318
2	90%	1,261	14,940
3	95%	584	6,984

15.3.2 THREE DIFFERENT SCENARIOS OF THE ABM

After modeling the base model, three different scenarios of IoT levels were analyzed in this case study. IoT technology focused not only on monitoring the condition of the refrigerator, but also on decreasing the failure rate. Scenario 1 simulates the behaviors of the refrigerator when the failure rate is reduced by 85%. Scenario 2 and 3 simulate the refrigerator behaviors when the failure rates are reduced by 90% and 95%, respectively. All the ABM results of the three scenarios are summarized in Table 15.4.

15.3.3 CONCLUSION

ABM is built to simulate the failure behaviors of the refrigerator in a retail pharmacy facility. The maintenance activity with a control parameter of failure rate is modeled in different scenarios while the ordering from the pharmacies and the warehouse act as the primary active agents. Implementing IoT to monitor the refrigerator failure shows a significant cost reduction. ABM results of Scenario 1 of IoT shows a total cost reduction of 86%, while the total cost was reduced by 91% and 95.82% in Scenarios 2 and 3 respectively. A limitation of this study is that the scope was limited to 20 pharmacies and one warehouse. Future work must increase the scope of the study. Also, the return on investment (ROI) of implementing IoT technology was not calculated, which is an important factor that should be considered for future work. Finally, the base model of ABM was not validated. Future work must include the validation of ABM and must calculate the degree of variation from the actual system.

15.4 CASE STUDY 2

A medical center located in Saudi Arabia was chosen for this study. The facility has more than 300 beds and covers an area of approximately 60,000 meters square. It is a six-story building with 25 networking rooms. There are two types of rooms: a main distribution frame (MDF) room and intermediate distribution frame (IDF) rooms. The former is where the main network and data warehouse are located. The multiple IDF rooms on each floor are connected to the MDF room. Both types of rooms are under the supervision of the Information Technology department (IT); however, if maintenance is required, the technical team is called. The current maintenance cost is around US$160k. The IT department has a scheduled check visit for all 25 rooms within the facility. If the IT team finds a failure, it will be reported to the technical team but not until they come back to the office. Then, a ticket will be initiated, and the technical team is required to fix the failure.

The purpose of this study is to improve the existing situation by minimizing the failure rate using ABM. Hence, to minimize the yearly maintenance cost. The current checking process is done manually, and a failure is not noticed until an IT team member visits the room or a failure is reported. Therefore, IoT sensors are considered in the proposed solution because it will save travel time in reporting the fault, instead receiving live feedback from the system. Moreover, it will enhance the predictive maintenance or scheduled maintenance. ROI is calculated and is to be considered in the final recommendation.

15.4.1 Process Model

An initial model is created to capture the current process, and this is depicted in Figure 15.1. The simulation base model shows that the IT department checks rooms on a daily basis based on a maintenance schedule. If a failure is observed, the maintenance team is called. Otherwise, checked items are noted in the system and scheduled for the next respected maintenance cycle.

Table 15.5 shows the model's parameter from the year 2020.

15.4.2 Simulation Model

Agents are built in such a way that interaction between them are defined without the demonstration of the process flow. Multiple states for each agent are defined and the behavior of the entity is known. We used the software AnyLogic for this case study. Four agents are interacting in the model: hospital, IT department, networking rooms, and the maintenance team. The proposed solution aims to utilize IoT to enhance the maintenance procedure. The hospital agent is represented with a simple dashed square. There are six floors in the building. The IT department agent is represented by a personal figure. They are located on the 3rd floor. The statechart of the IT agent and the JAVA codes used on start-up are shown in Figure 15.2.

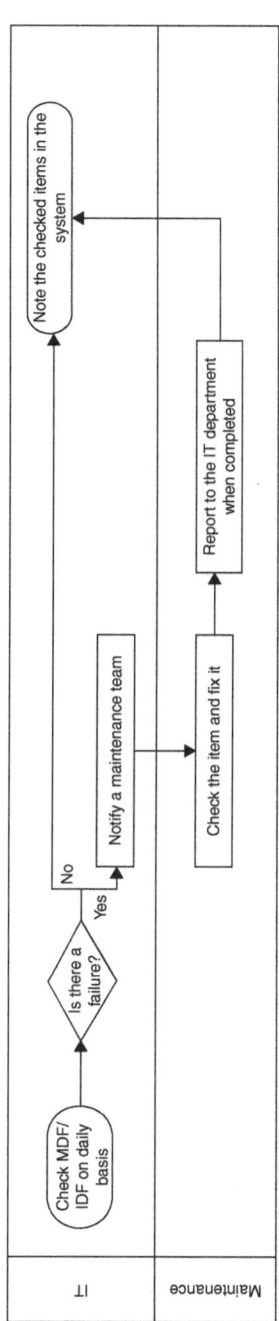

FIGURE 15.1 The current IT checking process

TABLE 15.5
Simulation Input Parameters

Parameter	Quantity	Unit
IT department	6	Technicians
Maintenance team	3	Technicians
Number of floors	6	Floors
Number of IDF/MDF Rooms	25	Rooms
Frequently of room checking	1	Weekly
Average time of completion	5	Days

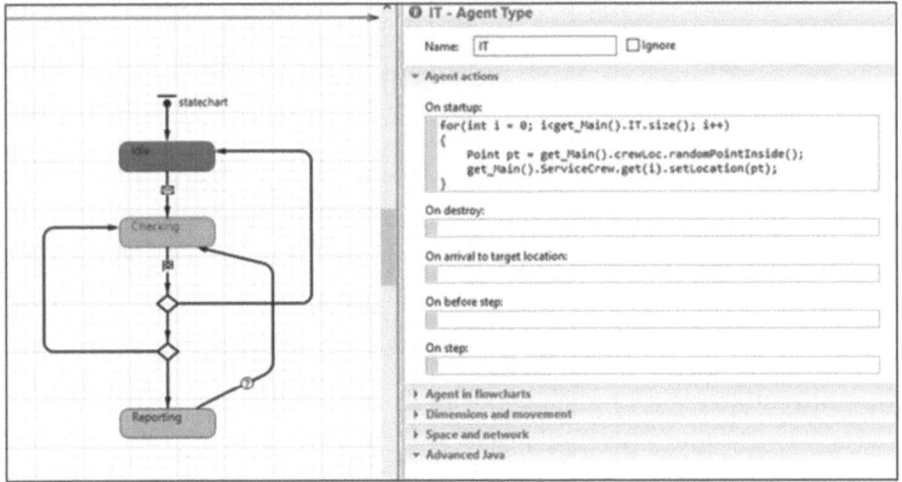

FIGURE 15.2   Statechart of the IT agent

The IT department can take different states as follows:

- Idle state.
- Based on a message received the agent moves to the respected room on the floor.
- The maintenance team moves to the respected room in case of the message received from the IT department.
- The IT team goes back to the home location (idle state) upon completing the scheduled visits.

The networking room agent indicates green color when it is working and red at the time a failure is found. Therefore, it has two states; either working or failed. The following states are considered for the maintenance team agent: the "idle" state is the default and where it begins. Upon receiving a message, the team becomes in a "movement" state, and once the agent arrives at the respected location the state changes to "fixing". When finished, the team goes back to the "idle" state.

15.4.2 ABM RESULTS

After running the simulation model for 1 year, the ABM shows that over 33% of devices in the medical center need maintenance with an approximate annual maintenance cost of US$160k. The management in the medical center wants to improve the failure rate of the devices by implementing an IoT project with an approximate cost of US$600k to monitor the devices and predict the failure.

15.4.4 THE RETURN ON INVESTMENT FOR THE PROJECT

ROI is calculated using the following equation:

$$\text{Discounted ROI} = \frac{(\text{PV cost} - \text{saving} - \text{PV Investment})}{\text{PV Investment}}$$

The ROI was determined by using different values of the failure rate reduction (85%, 90%, and 99% were used in the calculations). Table 15.6 illustrates the ROI values and the payback period (year 5) if the failure rate is reduced by 99%.

Table 15.7 shows a summary of ROI values associated with different rates of reduction. As is shown in the table, a positive ROI can be noticed at year 5 when the failure rate is reduced by 99%, while with a 90% rate of reduction the positive ROI is noticed two years later (year 7).

15.4.5 DISCUSSION

This research illustrates an important application of ABM. A case study is proposed to test the applicability of implementing an IoT project in a medical center. IoT

TABLE 15.6
ROI Values with 99% Reduction in the Failure Rate

Years	0	1	2	3	4	5
Cost saving		$149,011	$149,011	$149,011	$149,011	$149,011
PV (cost saving)		$141,242.65	$133,879.29	$126,899.80	$120,284.17	$114,013.44
Total PV (cost saving)		$141,242.65	$275,121.95	$402,021.75	$522,305.92	$636,319.36
Investment	$600k					
PV (investment)	$600k					
ROI		-76.46%	-54.15%	-33.00%	-12.95%	6.05%

Note: $ = US$.

TABLE 15.7
Summary of Different Failure Rates with Associated ROI Values

Failure Rate	ROI in Each Year							
Reduced By	1	2	3	4	5	6	7	8
85%	-95.24%	-90.73%	-86.45%	-82.40%	-78.56%	-74.91%	-71.46%	-68.19%
90%	-81.08%	-63.15%	-46.15%	-30.04%	-14.77%	-0.30%	13.42%	26.43%
99%	-76.46%	-54.15%	-33.00%	-12.95%	6.05%	24.06%	41.14%	57.32%

technology helps the facility to allow predictive maintenance features by monitoring the condition of the devices constantly. The maintenance activity of the IT rooms of the medical facility is modeled utilizing IoT to lower the overall cost. Finally, ROI is calculated to analyze the effects of employing IoT. The ABM results show that a positive ROI of 6.05% can be realized in year 5 with a 99% rate of failure reduction, and, when the failure rate is reduced by 90%, a positive ROI of 13.42% can be noticed in year 7.

REFERENCES

An, Gary, Qi Mi, Joyeeta Dutta-Moscato, et al. 2009. "Agent-based models in translational systems biology." *Wiley Interdisciplinary Reviews: Systems Biology and Medicine,* 1 (2): 159–171.

Bauer, Amy L., Catherine A. A. Beauchemin, and Alan S. Perelson 2009. "Agent-based modeling of host-pathogen systems: The successes and challenges." *Information Sciences,* 179 (10): 1379–1389.

Bruch, Elizabeth and Jon Atwell. 2015. "Agent-based models in empirical social research." *Sociological Methods & Research,* 44 (2): 186–221.

Chapizanis, Dimitris, Spyros Karakitsios, Alberto Gotti, et al. 2020. "Assessing personal exposure using agent-based modelling informed by sensors technology." *Environmental Research,* 192: 110141.

Dzaferagic, M., M. M. Butt, M. Murphy, et al. 2019. "Agent-Based Modelling Approach for Distributed Decision Support in an IoT Network." www.techrxiv.org/articles/preprint/ Agent-Based_Modelling_for_Distributed_Decision_Support_in_an_IoT_Network/ 11888871.

Grimm, Volker, Uta Berger, Finn Bastiansen, et al. 2006. "A standard protocol for describing individual-based and agent-based models." *Ecological Modelling,* 198 (1–2): 115–126.

Hinkelmann, F., D. Murrugarra, A. S. Jarrah, et al. 2010. "A mathematical framework for agent based models of complex biological networks." *Bulletin of Mathematical Biology,* 73: 1583–1602.

Macal, C. and M. North. 2009. "Agent-based Modeling and Simulation." Simulation Conference (WSC), Proceedings of the 2009 Winter, 13–16 Dec.

North, Michael J., Charles M. Macal, James St Aubin, et al. 2010. "Multiscale agent-based consumer market modeling." *Complexity,* 15 (5): 37–47.

Perera, Charith, Chi Harold Liu, and Srimal Jayawardena. 2015. "The emerging Internet of Things marketplace from an industrial perspective: A survey." *IEEE Transactions on Emerging Topics in Computing,* 3 (4): 585–598.

Ray, P. P. 2018. "A survey on Internet of Things architectures." *Journal of King Saud University – Computer and Information Sciences,* 30 (3): 291–319.

Rockett, Rebecca J., Alicia Arnott, Connie Lam, et al. 2020. "Revealing COVID-19 transmission in Australia by SARS-CoV-2 genome sequencing and agent-based modeling." *Nature Medicine,* 26 (9):1398–1404.

Samanidou, Egle, Elmar Zschischang, Dietrich Stauffer, et al. 2006. "Microscopic Models of Financial Markets." www.econstor.eu/bitstream/10419/3925/1/EWP-2006-15.pdf.

Savaglio, Claudio, Giancarlo Fortino, Maria Ganzha, et al. 2018. "Agent-based computing in the Internet of Things: A survey." *Studies in Computational Intelligence,* 737: 307–320.

Savaglio, Claudio, Maria Ganzha, Marcin Paprzycki, et al. 2020. "Agent-based Internet of Things: State-of-the-art and research challenges." *Future Generation Computer Systems,* 102: 1038–1053.

Silva, Petrônio C. L., Paulo V. C. Batista, Hélder S. Lima, et al. 2020. "COVID-ABS: An agent-based model of COVID-19 epidemic to simulate health and economic effects of social distancing interventions." *Chaos, Solitons & Fractals,* 139: 110088.

Whitmore, A., A. Agarwal, and L. Xu. 2015. "The Internet of Things – A survey of topics and trends." *Information Systems Frontiers,* 17 (2): 261–274.

Index